GEOMAGNETISM IN MARINE GEOLOGY

FURTHER TITLES IN THIS SERIES

Elsevier Oceanography Series, 6

GEOMAGNETISM IN MARINE GEOLOGY

by

VICTOR VACQUIER

Professor of Geophysics
University of California, San Diego, California;
Marine Physical Laboratory, Scripps Institution of Oceanography,
San Diego, California

ELSEVIER PUBLISHING COMPANY   *Amsterdam - London - New York 1972*

ELSEVIER PUBLISHING COMPANY
335 Jan van Galenstraat
P.O. Box 211, Amsterdam, The Netherlands

AMERICAN ELSEVIER PUBLISHING COMPANY, INC.
52 Vanderbilt Avenue
New York, New York 10017

Library of Congress Card Number: 78-190683

ISBN: 0-444-41001-5

With 149 illustrations and 6 tables

Printed in The Netherlands

PREFACE

In the past decade magnetic surveys of the oceans, combined with results from other branches of geophysics and geology have provided data for the construction of a plausible scheme for the evolution of the present distribution of land and oceans on the earth. According to this scheme, the surface of the globe is divided into roughly ten major rigid plates presently moving with respect to each other in response to unknown forces deep in the earth. Plates may consist of both continental and oceanic parts, their present boundaries being outlined by earthquakes.

Where two plates move away from each other new crust is formed at the rate of 1 to 12 cm/year, while when they move toward each other crust is being consumed often by underthrusting of a cold plate thus causing deep and shallow earthquakes. The most conspicuous spreading occurs in the ocean floor at the crest of oceanic ridges like the Mid-Atlantic Ridge where only shallow earthquakes occur because hot rock is close to the surface. As it cools, the new crust becomes magnetized in the direction of the magnetic field. When the field reverses its direction the rock is magnetized in the opposite way. Thus strips of positively and negatively magnetized rock form symmetrical bands on either side of the spreading ridges, which cause identical sequences of magnetic anomalies in widely different parts of the world. These properties of symmetry about the ridge axis and world-wide identity of the sequence of magnetic anomalies parallel to oceanic ridges establish beyond the shadow of a doubt that during the last 80 million years (m.y.) at least, the geomagnetic field can be regarded as due to a geocentric dipole lying along the earth's rotational axis, which from time to time rapidly changes from one polarity to the other, and that the banded magnetic anomalies constitute a record of sea-floor spreading at fairly uniform rate in the manner of magnetic tape in a sound recorder. The chronology of the geomagnetic reversals has been tied to radiometric ages of lava flows on land, paleomagnetic measurements on oceanic sediment cores and with paleontological ages from the JOIDES Deep Sea Drilling Program. Areas of the ocean floor several thousand kilometers in width have thus been dated to about 80 m.y. by magnetic anomalies and about 200 m.y. by the deep drilling. Magnetic anomalies in large areas of the oceans thus have recorded the motions of crustal plates in the geologic past, which along with paleomagnetic data lets us make likely paleogeographic reconstructions with respect to geographic north.

Although the relative movement of large crustal blocks was obviously accepted by Wegener (1929), Du Toit (1937) and other advocates of continental drift it was not until the delineation of the world-wide mid-ocean ridge system, that a reasonable place for crustal genesis in the wake of drifting continents could be identified (Heezen, 1962). Hess (1962) and Dietz (1962) proposed that continental drift was driven by convection of mantle rock rising under the mid-ocean ridges and sinking at deep-sea trenches and mountain ranges revived interest in the earlier speculations on mantle convection (Vening Meinesz, 1962; Griggs, 1939). The discovery of lineated magnetic anomalies by Mason and Raff (1961), and the correlation of those anomalies across fracture zones over distances of 1,400 km (Vacquier et al., 1961) demonstrated that the oceanic crust consists of rigid plates rather than a viscous medium provided for the flotation of continents. The idea that there is a connection between the growth of oceanic crust from spreading ridges and the generation of symmetrical patterns of lineated magnetic anomalies parallel to active oceanic ridges from geomagnetic field reversals was first published by Vine and Matthews (1963). After correlation of the magnetic anomaly pattern with the radiometric chronology of the geomagnetic field reversals in continental lava flows by Cox et al. (1964) and the discovery by Pitman and Heirtzler (1966) of the same sequence of anomalies from the Pacific–Antarctic Ridge, the Vine and Matthews method of detecting and dating sea-floor growth patterns was generally accepted. It stimulated the appearance of a massive literature on geomagnetic surveys in the oceans which this book attempts to review from the point of view of plate theory, as the latter is now providing the main incentive for making more magnetic surveys at sea.

This book was compiled for informing investigators in other branches of oceanography who often rub elbows with the marine geologists and geophysicists on oceanographic ships. It might be useful to students of marine geology and geophysics as a guide to literature. Perhaps teachers of earth science in secondary schools might use some parts of it as source material.

A large part of the material was assembled for a course of five lectures given at the Geological Institute of the U.S.S.R. Academy of Sciences in Moscow in May 1970. The inclusion of work published or about to be published since that date, so as to make the manuscript reasonably up to date, presented a difficult problem. In selecting published papers and the portions of them quoted in the book personal bias was inevitable, especially when I failed to agree with the author on some detail. Papers which in my opinion contain misleading conclusions and which are not discussed in the text were omitted from the bibliography. However, there are probably many others that were omitted simply by oversight. Sometimes for the sake of clarity and brevity bold statements are made in the text where a guarded discussion of alternatives would have been more appropriate.

This book was made possible by the cooperation of many colleagues who generously provided the figures quoted from their papers.

For critical review of the whole manuscript or of some of the chapters I am indebted to R.L. Fisher, M.D. Fuller, C.G.A. Harrison, B.C. Heezen, J.D. Mudie, and J.G. Sclater. R.L. Parker and Jean Francheteau contributed to Appendix II. J. Francheteau also assembled most of the tabulated data on the magnetism of ocean basalts. I also thank K.W. Milne for typing the manuscript, and Janice Anfossi for drafting some of the illustrations.

CONTENTS

CHAPTER 1

THE PHYSICS OF GEOMAGNETISM

MAGNETIC INTENSITY

When a ship tows a magnetometer, this instrument measures the magnitude but not
the direction of the magnetic field intensity. The magnetic field intensity is a vector
quantity best defined by the magnetic effect at the center of a circular loop of wire of
radius r carrying an electric current i as shown in Fig.1. The *magnetic field* $\vec{H}$, an abbrevia-

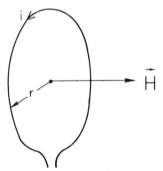

Fig.1. Definition of magnetic intensity.

tion for "magnetic field intensity", is a vector perpendicular to the plane of the current
loop. Its magnitude is:

$$\vec{H} = \frac{2\pi i}{r} \qquad \text{Oe (oersteds) or gauss, } \Gamma$$

where i is measured in electromagnetic units (1 e.m.u. = 10 A) and r is in cm. In
geomagnetism the gamma (1 γ = 10^{-5} gauss)[1] or the milligauss (= 10^{-3} Γ) are commonly
used. As we shall mention later on, the earth's magnetic field is generated by electric
currents circulating in the liquid iron core below 2,900 km from the earth's surface.
Because of this great depth, the so-called "permanent" geomagnetic field (to distinguish it
from the short-time varying part due to solar radiation and particle emission) is geographi-
cally smooth compared to the field anomalies arising from the magnetic floor of the
ocean which are superposed on it.

---

[1]In the new S.I. (Système International d'Unités) system of units 1 γ = 1 nanotesla.

INTENSITY OF MAGNETIZATION

Omitting for the present the details of rock magnetism, let us regard a bar of magnetized rock shown on Fig.2 as consisting of small elementary magnets generally oriented along the axis of the bar. The bar's total magnetism, called its *magnetic moment*, is designated on Fig.2 as the vector $\vec{M}$. The magnitude of $\vec{M}$ is defined as the turning effort or torque in dyne cm that the bar would experience when placed at right angles in a field of one gauss as shown on Fig.3. By convention the end of the bar that points northward

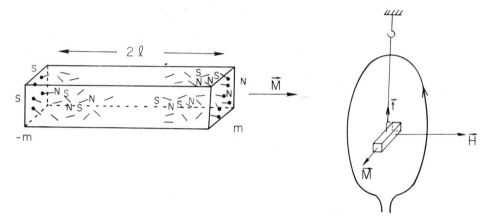

Fig.2. Diagrammatic representation of a uniformly magnetized bar. Only the elementary poles on the extremities produce effects at a distance. $\vec{M} = Jv = 2\,ml$ c.g.s.

Fig.3. Definition of magnetic moment. $\vec{H} = 1\ \Gamma . \vec{t} = 1$ dyne cm. $M = 1$ c.g.s.

is called the north-seeking pole. If one regards the earth as having a magnet at its center, the north pole of this magnet points toward the geographic South Pole. If the degree to which the material of the magnet is magnetized is uniform throughout its volume v, its intensity of magnetization. $J = \vec{M}/v$ electromagnetic units (e.m.u.) or e.g.s. units of magnetic moment per cm^3.

FIELD DUE TO A DIPOLE

In what follows field measurements will be compared with fields computed from simple models, the simplest one of which is the bar magnet. At a point located at a distance several times the length of the uniformly magnetized bar of Fig.2, the magnetic intensity due individual north and south poles of the elementary magnets inside the volume of the bar cancel one another except at the end surfaces. This is convenient because when one calculates the effect of a uniformly magnetized mass of rock, one can integrate the distribution of these fictitious poles populating the surface of the body which is in general a simpler task than carrying out an integration of the volume magnetization. This substitution is called Gauss' theorem in the theory of the Newtonian potential. If the linear dimensions of the ends of the magnetized bar are small compared to the distance r to the point P of Fig.4 where the field of the magnet is measured, the

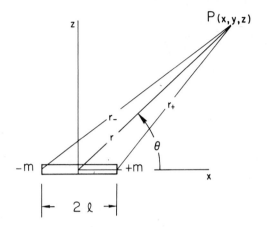

Fig.4. Schematic representation of a magnetic dipole. The y axis is directed into the paper.

product of the surface intensity of magnetization times the area can be regarded as a fictitious pole of strength m at one end and $-m$ at the other located at a distance l from the center of the bar. The magnitude of the magnetic moment of the magnet is then $\vec{M} = Jv = 2\,ml$. Now, when one applies Newton's inverse square law of attraction by assuming that the poles m and $-m$ are equal positive and negative masses, the correct functional form is derived for the magnetic field generated by the magnet in the space external to it. To verify this statement one can compare the field produced by the magnet to that of the current loop of Fig.1 which is equal to $\pi r^2 i$ at distances large compared to the size of the loop.

We shall derive the expression for the magnetic potential at the point P of a dipole represented by the magnet of Fig.4. Then the magnetic intensity in any direction can be gotten by differentiation of the potential. This simple calculation helps to understand quantitatively the magnetic effects produced by magnetized bodies. Referring to Fig.4, the potential U at P $(x,\ y,\ z)$ is the sum of the contribution of the positive and negative poles of the magnet:

$$U = \frac{m}{r_+} - \frac{m}{r_-} = m\left(\frac{1}{r_+} - \frac{1}{r_-}\right)$$

where:

$$r_+ = [(x-l)^2 + y^2 + z^2]^{1/2}$$

and:

$$r_- = [(x+l)^2 + y^2 + z^2]^{1/2}$$

Since l is much smaller than x, y, and z, we neglect the quantities containing l^2:

$$r_+ = (x^2 - 2xl + y^2 + z^2)^{1/2} = (r^2 - 2xl)^{1/2} = r \left(1 - \frac{2xl}{r^2}\right)^{1/2}$$

$$r_- = (x^2 + 2xl + y^2 + z^2)^{1/2} = (r^2 + 2xl)^{1/2} = r \left(1 + \frac{2xl}{r^2}\right)^{1/2}$$

The approximate expression for the potential becomes:

$$U = \frac{m}{r} \left[\left(1 - \frac{2xl}{r^2}\right)^{-1/2} - \left(1 + \frac{2xl}{r^2}\right)^{-1/2} \right]$$

Now expand the inner brackets in a power series and again the square and higher power terms of $2xl/r^2$ can be neglected because it is much smaller than unity. The power series expansion of:

$$(1 \pm a)^{-1/2} = 1 \pm \frac{1}{2}a + \frac{1.3}{2.4}a^2 \pm \; \ldots$$

which when substituted into the formula for the potential gives:

$$U(x,y,z) = 2lm \frac{x}{r^3} = \vec{M} \frac{x}{r^3}$$

Where $\vec{M}$ is the magnetic moment of the magnet. If θ is the angle between the magnetic moment and r, the potential can also be written:

$$U(P) = \vec{M} \frac{\cos \theta}{r^2} \tag{1}$$

which is independent of the coordinate system. The derived expressions are absolutely accurate only in the limit when $l \to 0$. Simultaneously, the pole strength m can be increased so as to keep $\vec{M} = 2\,lm$ constant. This idealized magnet is called a dipole.

The negative derivative of the potential with respect to a direction yields the magnetic field intensity in this direction. As previously stated, this is a direct consequence of the inverse square law of gravitational attraction that was assumed to be valid for positive and negative poles of the magnet of Fig.2 by thinking of them as concentrated positive and negative masses. The magnetic intensity in the x direction:

$$\vec{H}_x = -\frac{\partial U}{\partial x} = \frac{\vec{M}}{r^3}\left(\frac{3x^2}{r^2} - 1\right) \tag{2}$$

Two particular positions are useful to remember. One is on the x axis with coordinates $x = r$, $y = 0$, $z = 0$:

$$\vec{H}_x\,(x = r, y = 0, z = 0) = \frac{2\vec{M}}{r^3} \tag{3}$$

The other is on the z, y plane:

$$\vec{H}_x\,(x = 0, \sqrt{y^2 + z^2} = r) = -\frac{\vec{M}}{r^3} \tag{4}$$

INCLINATION OF THE GEOMAGNETIC FIELD

An elementary application of the preceding formulas to the earth's magnetism comes to mind. As a first approximation the earth's magnetic field is representable by a magnetic dipole at the center of the earth directed to the South Geographic Pole. It is called the "geocentric dipole", the field of which at the surface of the earth is given by the first term of the spherical harmonic representation of the earth's permanent magnetic field in terms of geographic latitude and longitude. On Fig.5 the total magnetic intensity

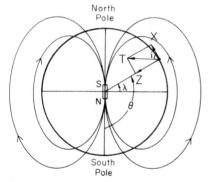

Fig.5. Inclination of magnetic field of geocentric dipole. See eq.5 and 6. Tan i = 2 tan λ . θ = 90° + λ.

T at a point of latitude λ on the earth can be obtained from the vector sum of the north component X and the vertical component Z, which are obtained from the negative derivatives of eq.1 with respect to the distance north $rd\theta$ and r directed downward. The latitude λ is related to θ by θ = (90° + λ) so that cos θ = −sin λ. Also the northward direction is opposite to the direction of the magnetic dipole. Keeping these directions straight, we get for the northward component:

$$X = \frac{-1}{r}\frac{\partial U}{\partial \theta} = \frac{\vec{M}}{r^3}\cos \lambda \tag{5}$$

and for the vertical component:

$$Z = \frac{-\partial U}{\partial r} = \frac{2\vec{M}}{r^3} \sin \lambda \qquad\qquad\qquad (6)$$

Eq.5 and 6 confirm the observation that the geomagnetic field is twice stronger at the poles than at the equator. From the value of the observed field, which is about 0.3 G at the equator one can compute the magnitude of the fictitious geocentric dipole to be about $8 \cdot 10^{25}$ c.g.s. units.

A SIMPLE TWO-DIMENSIONAL MODEL FOR ROUGH ESTIMATES

The geomagnetic anomalies measured from ships and airplanes come from magnetized igneous rock within the crust of the earth. The larger part of this magnetization has been acquired as the rocks cooled down in the presence of the earth's field from the Curie temperature, about 600°C or cooler depending on the composition of the magnetic minerals. Above the Curie temperature the rocks are effectively non-magnetic. From what is known about the rise of temperature with depth, the magnetic anomalies at sea can not come from a depth much greater than 20 km, and from other considerations it is likely that the magnetic layer is much thinner. Present evidence indicates that basalt is the only rock type one needs to consider as the source of magnetic anomalies, and that it gets magnetized either in the present direction of the earth's field or in the opposite direction depending on whether the geomagnetic field was normal or reversed at the time the rock cooled down from the Curie temperature. Identical sequences of magnetic anomalies parallel to spreading ridges in different oceans prove beyond the shadow of a doubt that

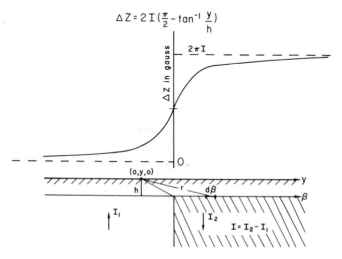

Fig.6. Vertical magnetic intensity from a vertically magnetized quarter space. When I is in e.m.u. and the angle in radians, ΔZ comes out in gauss. The x axis is directed into the paper.

the polarity of the earth's field flips $180°$ at irregular intervals. The calculation of the magnetic intensity produced by long strips constant in thickness of alternate polarity is simple because the effects of the individual strips just add up. Another simplification results if one chooses a ridge oriented north–south in middle latitude, for then only the vertical magnetization in the rock and the vertical magnetic intensity at the ship enter the picture.

In Fig.6 let h represent the depth of the ocean. The vertical, north–south plane through the origin separates two quarter spaces which are oppositely magnetized, the difference of their magnetizations being I, which can be assigned to the righthand space since the anomaly arises only from the difference. In reality, the magnetized bodies possess a bottom surface, but its effect can be neglected if their depth is very great compared to the water depth h. The coordinates of the ship are $(0, y, 0)$. The coordinate y is directed east, x into the paper. The element of surface on the righthand quarter space $d\alpha\, d\beta$ at α and β. This surface is populated with a surface density of south poles. The vertical magnetic intensity at the ship is:

$$\Delta Z = \int_{-\infty}^{\infty} \int_{-\infty}^{\infty} \frac{I d\beta\, d\alpha}{r^2} \cos\theta$$

Now $\cos\theta = h/r$, so:

$$\Delta Z = \int_{-\infty}^{\infty} \int_{-\infty}^{\infty} I \frac{h}{r^3} d\beta\, d\alpha$$

where $r = [(\alpha - x)^2 + (\beta - y)^2 + h^2]^{1/2}$. Integrating first with respect to α we have left:

$$\Delta Z = 2I \int_{-\infty}^{\infty} \frac{h\, d\beta}{(\beta - y)^2 + h^2} = 2I \tan^{-1} \frac{\beta - y}{h} \Big|_{-\infty}^{\infty} = 2I \left[\frac{\pi}{2} + \tan^{-1} \frac{y}{h} \right] \tag{7}$$

As $y \to -\infty$, $\tan^{-1} \dfrac{y}{h} \to -\dfrac{\pi}{2}$, and $\Delta Z \to 0$

As $y \to +\infty$, $\tan^{-1} \dfrac{y}{h} \to \dfrac{\pi}{2}$, and $\Delta Z = 2\pi I$

which is the maximum anomaly for a given intensity of magnetization.

In practice the bottom can not be neglected. To provide for it, one simply subtracts the result of the previous calculation for a deeper depth h' which then gives the vertical intensity of a semi-infinite slab or fault:

$$\Delta Z = 2I \left(\tan^{-1} \frac{y}{h} - \tan^{-1} \frac{y}{h'} \right) \tag{8}$$

The two terms in the bracket are the angles expressed in radians which are subtracted at the magnetometer by the upper and lower surfaces of the slab shown on Fig.7.

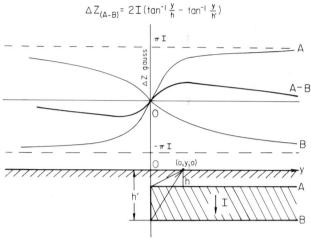

$$\Delta Z_{(A-B)} = 2I(\tan^{-1}\tfrac{y}{h} - \tan^{-1}\tfrac{y}{h'})$$

Fig.7. Vertical magnetic intensity from a vertically magnetized semi-infinite slab or a fault is obtained by subtracting the contribution of a quarter space whose horizontal surface is at a deeper depth h' from the curve of Fig.6.

If one wishes to limit the slab in the y direction to get the effect of a slab which is $2b$ wide (Fig.8), one combines the solutions in which y is replaced by $y - b$ and $y + b$.

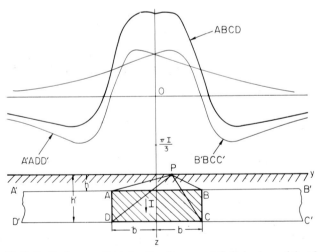

Fig.8. Vertical magnetic intensity from an infinitely long slab of rectangular cross-section $ABCD$ obtained by adding the effects of two negatively magnetized faults of Fig.7 separated by a distance $2b$. Note that as the body is widened the amplitude of the anomaly over the center of the body approaches zero. The ordinates have been stretched 3 x compared to the previous two figures as indicated by the ticks on the vertical axis. The difference $<APB - <DBP = 0.9193$ radians, giving a value of 1839 γ for the anomalous field at point P for a magnetization of 0.01 e.m.u.

Geometrically, ΔZ is again proportional to the difference of the angle subtended by the top and by bottom of the body. For example, the block of Fig.8 for $I = 0.01$ e.m.u. gives a maximum positive anomaly of 1,840 γ at $y = h$, which can be checked by subtracting the angle *DPC* from the angle *APB*.

If the body is limited in the x direction, the vertical intensity that it gives is proportional to the difference of the solid angles (instead of plane angles) subtended by its top and bottom surfaces.

Model calculations for the general case in two dimensions when the strike of the lineations is not northward are more complex and are treated in Appendix 2.

CHAPTER 2

MAGNETIC PROPERTIES OF OCEANIC BASALTS

DIAMAGNETISM AND PARAMAGNETISM

If a substance acquires an intensity of magnetization J (c.g.s./cm^3) in a field of H (gauss) the volume magnetic susceptibility $k = J/H$, while the mass susceptibility $\chi = J/\rho H$ where ρ (g/cm^3) is the density. A substance is called diamagnetic if χ and k are negative, whereas in the case of paramagnetic and ferromagnetic substances they are positive.

In diamagnetic substances the net magnetic moment of atoms is zero. The diamagnetism arises from precession of the electron cloud about the external field as the electrons rotate about the nucleus, and is roughly proportional to the number of electrons. Diamagnetism is extremely feeble. For example $k = 13 \cdot 10^{-6}$ for bismuth.

Although oceanic basalts and sediments are ferromagnetic, an elementary discussion of paramagnetism is needed for understanding their magnetic properties. Substances having an odd number of electrons are paramagnetic, that is, their magnetic susceptibility is positive because the net magnetic moment of the atoms is not zero, as in diagmagnetic substances. Paramagnetism is generally ten or more times stronger than diamagnetism. The magnetic properties of minerals can be traced to atoms with odd numbers of electrons in their electron shells so that some electron spins are not cancelled. In a magnetic field the atoms of a paramagnetic substance tend to line up their magnetic moments parallel to the field against the disordering influence of thermal agitation which is much greater at ordinary temperatures. A distinctive characteristic of a paramagnetic substance is the linear dependence of its susceptibility on the reciprocal of the absolute temperature given in gases by $\chi = C/T$. This is called Curie's law which links magnetism to the kinetic theory of gases. Curie's constant C is proportional to the square of the molecular magnetic moment. In solids, because of interaction between atoms, the mass susceptibility is expressed as $\chi = C/(T - \theta)$, which is valid only when $T < \theta$. θ is called the Curie Temperature.

FERROMAGNETISM

Below this temperature some paramagnetic metals, especially iron, nickel and cobalt as well as some of their compounds are much more magnetic than normal paramagnetic substances. For example, in a field of 10 Γ, the mass susceptibility is $-0.8 \cdot 10^{-6}$ for antimony, $65 \cdot 10^{-6}$ for cobalt sulfate and 200 for pure iron. These solids are called ferromagnetic. Their unusually strong magnetic properties below the Curie temperature arise from interaction between atoms in the crystal. Above the Curie temperature they become paramagnetic. A ferromagnetic substance consists of discrete regions containing 10^{10} to 10^{15} atoms called "domains" in which the magnetic moments of the atoms are all parallel to each other. Since in a polycrystalline substance individual crystals containing magnetic domains are randomly oriented, the bulk magnetization is zero in the unmagnetized state. It is said that the domains possess "spontaneous magnetization" due to interatomic interaction. This spontaneous magnetization happens in a field-free space, and its magnitude in each domain is the maximum for the material.

In a field-free space, as the material cools from the Curie temperature, the domains become spontaneously magnetized along a preferred crystallographic axis, the axis of "easy" magnetization in which the magnetic energy is minimum and the magnetic interaction between the atoms of the lattice is maximum. Even carefully grown single crystals of pure iron consist of many domains caused in part by slight imperfections and occasional impurities causing strains in the crystal lattice. The size of magnetic domains has been studied by observing under a microscope on polished specimens of ferromagnetic metals the aggregation of colloidal ferromagnetic particles at the domain boundaries where the magnetic field gradient is especially strong. The Barkhausen Effect is also used for estimating the volume of magnetic domains. This experiment consists of slowly magnetizing a specimen in a pick-up coil. The latter picks up discrete bursts of energy which presumably arise from individual domain walls shifting to another stable position. The magnetization of a ferromagnetic substance by an external field consists of three parts as illustrated by the diagram of Fig.9. At first the domains with magnetizations that have a component along the applied field reversibly grow in size at the expense of the

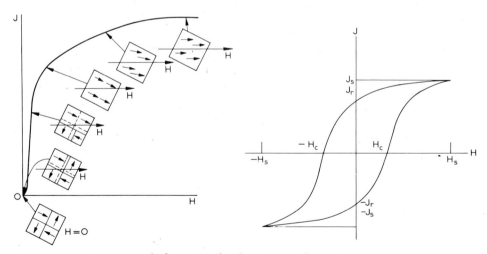

Fig.9. Magnetization process of a ferromagnetic substance according to domain theory as illustrated by Irving (1964).

Fig.10. Hysteresis of cyclic magnetization of a ferromagnetic substance. (After Irving, 1964.)

domains directed contrary to that field. Then some domain walls suddenly jump to another stable position which increases the total magnetic moment in the direction of the applied field, and finally the atoms in all domains continue to be reversibly deflected by the applied field from their stable orientation. This deflection is, of course, present for all values of the external field. Because of the energy barriers between the stable positions of the domain walls, the magnetization of ferromagnetics goes through a hysteresis cycle in an alternating field as shown on Fig.10. The specimen is magnetized to saturation J_s at a field H_s. Then, as the field is reduced to zero, some of the domain walls remain in their newly acquired position giving the specimen a remanent magnetization J_r. To wipe out this remanent magnetization a negative field $-H_c$ called the "coercive force" has to be applied.

ANTI-FERROMAGNETISM AND FERRIMAGNETISM

In the case of chemical compounds, the ordering of atoms in the domains can be of three different types schematically illustrated in Fig.11. So far we have considered only the pure ferromagnetic kind (Fig.11A) in which the magnetic moments of the atoms of a domain are all parallel to each other. In anti-ferromagnetic substances (Fig.11B) two equal sub-lattices have oppositely directed moments so that normally the substance is not ferromagnetic. However, as the temperature is raised one sub-lattice becomes disordered first, causing a sharp increase of magnetic susceptibility. As the temperature is increased some more, the susceptibility drops again, the material becoming paramagnetic. The temperature at which the susceptibility has a sharp peak is called the Néel temperature. Domains in ferrimagnetic crystals of Fig.11C consist of two antiparallel sub-lattices of unequal moment giving the substance generally ferromagnetic properties but producing peculiar shapes of the magnetization vs. temperature curves.

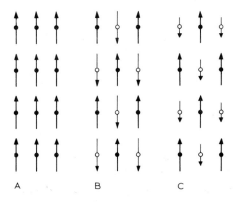

A   B   C

Fig.11. Diagrammatic representation of ferromagnetic (A), anti-ferromagnetic (B) and ferrimagnetic (C) crystals. (After Nagata, 1961.)

COMPOSITION OF FERROMAGNETIC ROCK-FORMING MINERALS

Magnetic properties of rocks depend on their composition as well as on their history. The only important ferromagnetic minerals in ocean basalts are titaniferous magnetite and titaniferous maghemite, which are solid solutions of oxides of iron and titanium. Two families of minerals are shown on Fig.12: the magnetite-ulvospinel series and the

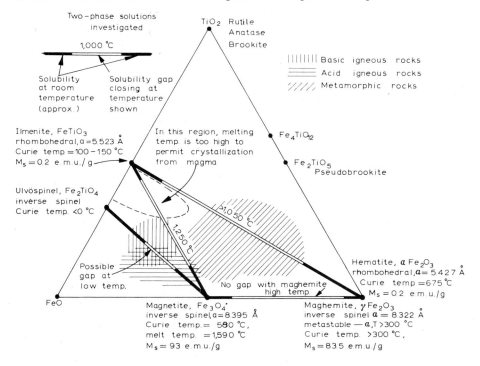

Fig.12. Mineralogical and magnetic data concerning the system FeO, TiO_2, Fe_2O_3. (From Grant and West, 1965. Used with permission of McGraw-Hill Book Company and the authors.)

hematite—ilmenite series. The magnetite-ulvospinel minerals are ferrimagnetic whereas the hematite—ilmenite series is much less magnetic being essentially anti-ferromagnetic with a small defect making them weakly ferromagnetic. In basalts the magnetite occurs both in the ground mass and as small phenocrysts.

Figs.13 and 14 give the dependence of the Curie point on the composition for both families of minerals. Although it would be reasonable to expect that one could predict the value of the Curie point from the quantitative chemical analysis of the oxides, it turns out that the actual Curie temperatures are usually much higher than the predicted ones. Even when the magnetic minerals are separated from the silicates, microscopic examination reveals exolved plates of ilmenite within the titanomagnetite crystals which shifts the Fe/Ti ratio toward higher Curie temperatures (Ade-Hall, 1963). It is thus impossible to predict magnetic properties of rocks from chemical analysis.

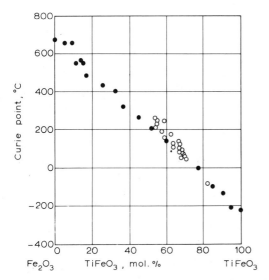

Fig.13. Curie points in the ilmenite-hematite series. Circles are natural minerals, dots synthetic minerals. Redrawn from Nagata and Akimoto (1961, quoted in Irving, 1964) with the kind permission of Maruzen, Tokyo, and the authors. (After Irving, 1964.)

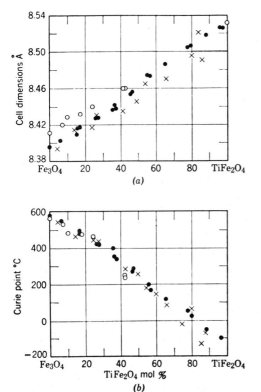

Fig.14. Variation of size of unit cell (A) and Curie temperature (B) in the magnetite-ulvospinel solid solution series. Values from synthetic specimens. Dots are values from Akimoto, Katsura, and Yoshida (1957, quoted in Irving, 1964), crosses from Kawai (1959, quoted in Irving, 1964), and circles from Pouillard (see Nichols, 1955, quoted in Irving, 1964). (After Irving, 1964.)

EFFECTS OF DOMAIN AND GRAIN SIZE

As previously mentioned, magnetic anomalies in oceanic areas come largely from permanently magnetized rock bodies rather than from magnetization induced by the presently existing geomagnetic field. This magnetized state possesses a higher internal energy than the unmagnetized state, and it is natural to inquire whether remanent magnetism decays with time which may be of the order of 100 million years (m.y.) for rocks.

Néel (1955) has devised a theory from which the stability of the magnetic moment of a single domain particle of given size can be calculated as a function of time for a given temperature. Laboratory experiments on igneous rocks and synthetic minerals closely agree with this theory, making it likely that a large and especially the stable part of the remanent magnetism of igneous rocks is carried by small single domain grains dispersed in a nearly nonmagnetic matrix. Consider the hysteresis loop of a single isolated domain, and let the external applied field be parallel to the spontaneous magnetization of the domain as shown in Fig.15A. The loop is rectangular because only two magnetized states $\pm J_s$ are possible. When the applied field is perpendicular to the direction of spontaneous magnetization, the induced magnetization parallel to the applied field is reversible (Fig.15B). The hysteresis loop of a material in which domains of some size are randomly dispersed would look like Fig.15C, a fair imitation of a hysteresis loop.

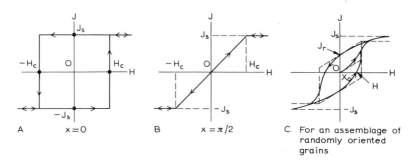

Fig.15. Hysteresis loop of single domain particles predicted from Néel's theory. (After Nagata, 1961.)

The atoms of a domain are kept aligned by forces proportional to the product of its volume v and the magnetocrystalline anisotropy K_u, the latter being the difference in energy that is required to magnetize the crystal in the easy direction of magnetization compared to the other crystal axes. This directive force is fought by thermal agitation kT, where k is Boltzmann's constant $1.380 \cdot 10^{-16}$ erg/°K and T the absolute temperature. For a grain to be ferromagnetic, the ratio $K_u v/kT$ should be above a certain critical value. At slightly below the Curie temperature, statistical fluctuations in thermal agitation will reduce a remanent magnetization exponentially with time as $J_r = J_0 \exp(-t/\tau_0)$, where J_0 is the initial value of J at zero time and τ_0 is the relaxation time of the grains. Néel derived the expression for τ_0 (Nagata, 1961, p.23) the principal characteristic of which is the exponential dependence of τ_0 on v/T. Table I gives values of τ_0 for spherical particles

TABLE I
Dependence of the relaxation time τ_0 on v/T or the radius at $T = 300°K$ for magnetite (spherical shape)[1] (After Nagata, 1961)

τ_0 (sec)	10^{-1}	10	10^2	10^3	10^5	10^7	10^9	10^{15}	100^{100}
$v/T(10^{-19}$ cm^3/deg.)	2.5	3.2	3.5	3.8	4.4	5.1	5.7	7.6	34.6
R (Å) at $T=300°K$	260	280	290	300	320	330	340	380	630

[1] Anisotropy is assumed to be exclusively of magnetocrystalline origin.

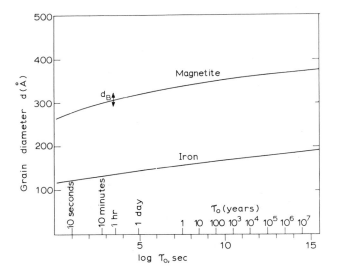

Fig.16. Dependence of relaxation time τ_0 on particle diameter at room temperatures for iron and magnetite. Iron data from Néel (1955). Magnetite data from Nagata and Akimoto (1961, quoted in Irving, 1964) who assume anisotropy to be entirely magnetocrystalline. The arrows indicate the range of critical blocking diameters d_B. (From Irving, 1964.)

of magnetite. The materials for which τ_0 is shorter than the time it takes to measure J_r, as illustrated at the left edge of Fig.16 for iron and magnetite, are called super-paramagnetic. The coercive force depends on particle size. A solid dispersion of single domain grains has maximum coercive force, and we have just seen that with decreasing grain size the material becomes super-paramagnetic. However, where grains are large enough to consist of several domains, the coercive force drops again and with increasing particle size approaches asymptotically the value for the bulk material as shown schematically in Fig.17 along with the initial susceptibility which has a sharp maximum where the domains are in the super-paramagnetic state. The drop of the coercive force may be interpreted as caused by a hysteresis of the motion of domain walls in the multi-domain structure.

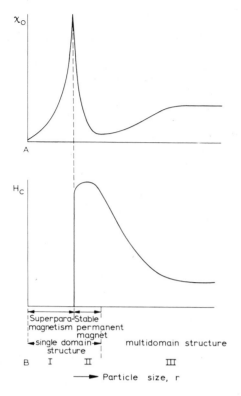

Fig. 17. Initial susceptibility (A) and coercive force (B) of magnetite as dependent on particle size. (After Nagata, 1961.)

THERMOREMANENT MAGNETIZATION

The remanent magnetization of igneous rocks is generally from 5 to 100 times the magnetization that is induced in them by the present earth's field. Their coercive force is usually several hundred gauss, which is comparable to some permanent magnet steels (see Fig. 21). These fortunate properties can be explained by the single-domain particle theory of Néel mentioned in the preceding section. As the single-domain particles cool through the *blocking temperature* at which they pass from the super-paramagnetic state to the ferromagnetic one, their coercive force is extremely small, and the weak geomagnetic field can magnetize them much more effectively than at an even slightly lower temperature. At fields as low as the earth's this thermoremanent magnetization is linearly proportional to the magnitude of the field. Because the domains have different volumes, there is a range of temperature in which a rock acquires its thermoremanent magnetization. Most of the TRM, as it is called, is acquired within 100° of the Curie temperature as illustrated in Fig. 18. The figure also shows that partial TRM's are approximately additive, that is, the TRM from 600–300°C equals the sum of the TRM's between 600–500°C and

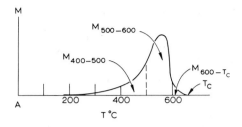

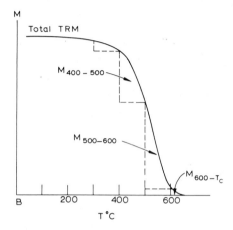

Fig.18. The acquisition of TRM. In these idealized curves the partial TRM's are plotted, (A) separately, and (B) cumulatively, as a function of temperature. Modified from Cox and Doell (1960). (After Irving, 1964.)

500–300°C. Experiments on rocks and synthetic minerals are consistent with the supposition that a considerable fraction of the magnetic minerals in basalts is a dispersion of single-domain particles which produces the TRM, which is extremely stable because at room temperature, the coercive force of the single-domain particles is many times greater than any magnetizing field it might be subjected to. However, igneous rocks do contain particles large enough to consist of many domains and this gave rise to several proposals for explaining how multi-domain particles can behave as single-domain ones. Stacey (1969, p.162) has advanced the explanation that if a particle consists of only a few domains, the stable positions of the domain walls are not necessarily those that would produce zero over-all magnetization of the particle. Strangway et al. (1968) provide a mechanism for this idea. They propose that titaniferous magnetic crystals contain exsolved laminas of ilmenite which effectively divide the magnetic grain into regions small enough to act as single domains. Verhoogen (1959) has proposed that TRM comes primarily from dislocations which behave like single domains because of being isolated from interaction with other parts of the ferromagnetic grain by stressed regions. Such magnetization would have greater stability than IRM which might arise in greater part from hysteresis of the motion of domain walls (Ozima and Ozima, 1965). An interesting property of TRM is that if it is acquired in a weak field like the field of the earth, on heating it decreases appreciably only within the immediate neighborhood of the Curie point.

Perhaps we just imagine the need for an explanation of TRM in multi-domain particles, and that the bulk of the magnetically hard portion of the NRM resides in single-domain grains as experiment and theory are telling us (Lowrie and Fuller, 1971).

THERMOMAGNETIC PROPERTIES OF OCEANIC BASALTS

Reversible and irreversible changes in magnetic properties of basalts caused by various kinds of heat treatment have been used to identify the magnetic minerals they contain. The magnetic minerals are also observed before and after treatment by X-rays, optical and electron microscopy, etc. Specimens of synthetic mineral dispersions are subjected to the

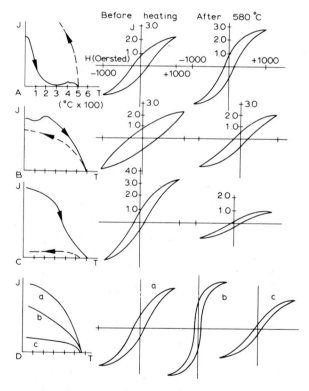

Fig.19. Illustration of thermomagnetic characteristics of the three classes of ocean basalts studied by Wasilewski. At left are Curie point analyses, at right are magnetization curves before and after heating. In D three different examples are illustrated, since the thermomagnetic behavior of class III curves is reversible.

A. Class I; sample CH21.001; Mid-Atlantic Ridge. B. Subclass IIA; sample D8.02; Puerto Rico Trench. C. Subclass IIB; sample D8.015; Puerto Rico Trench. D. Class III; three samples (a, b, and c). (After Wasilewski, 1968.)

same heat treatment and examination. It was found that for two species of basalt the magnetization acquired in a strong field and the Curie point increase irreversibly after heating. In some basalts studied by Wasilewski (1968) (Fig.19) a homogeneous titano-magnetite rich in titanium splits upon heating into a titanomagnetite poorer in titanium and ilmenite. Ozima and Larson (1970) and Ozima and Ozima (1971) obtained the same magnetic behavior in a basalt containing highly oxidized titanomaghemite which got transformed to magnetite and hemo-ilmenite and/or pseudo-brookite. Wasilewski (1968) also found another specie of basalts (class II, Fig.19B,C) for which the magnetization decreased while the Curie point remained unchanged. The titomagnetite in this basalt is changed presumably into a defect spinel which is less magnetic than the original mineral. Fig.19D shows the thermomagnetic curves of another kind of submarine basalt the magnetic properties of which are reversible with respect to heating and cooling.

Low-temperature oxidation of the original titanomagnetite of a freshly-extruded basalt from a ridge crest reduces its NRM, which would explain the observation that in computing magnetic anomalies arising from sea-floor spreading one often has to assign a larger value of magnetization to the central block to match the observed magnetic anomaly.

REMOVAL OF UNSTABLE MAGNETIZATION

The presence of more than one magnetic mineral in rock may be found by steps in the variation of magnetic properties such as susceptibility or remanent magnetization with temperature (Fig.20) provided no irreversible mineralogical alterations are caused by

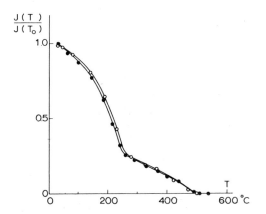

Fig.20. Thermomagnetic curve of mineral having two components with different Curie points. Groundmass ferromagnetic mineral in hypersthene basalt, 1950 lava of Mihara, Osima. H_{ex} = 2,400 Oe. (From Nagata, 1961.)

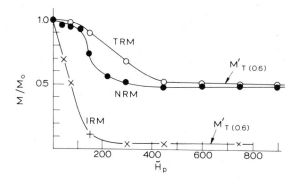

Fig.21. Comparison of alternating field demagnetization of IRM, NRM, and total TRM in the same sample of trachyte from the Aden volcanics. Normalized intensity is plotted against peak field strength. (After Irving, 1964.)

heating. This can also be done by demagnetization in an a.c. field illustrated in Fig.21 which shows that the total magnetization consists of a soft component which is destroyed below 200 Oe and an extremely hard one which is unaffected by a.c. fields greater than 900 Oe. The hard component of the natural remanent moment (NRM) is equal to the experimentally produced thermoremanent magnetization (TRM). The soft component may originate from a mineral having a lower coercive force, or from the action of the earth's field at a temperature considerably below the Curie point called viscous remanent magnetization (VRM).

MAGNETIC VISCOSITY

Although one can visualize VRM as being due to super-paramagnetic grains or to displacement of domain walls stimulated by thermal agitation, these mechanisms fail to explain all the observed properties of VRM. This complex and incompletely understood subject is comprehensively reviewed up to 1965 by Sholpo (1967). For our practical purposes it is sufficient to evaluate the possible influence of VRM on the interpretation of marine geomagnetic surveys from its observed properties. If the directions of TRM and VRM are different, NRM, which is their vector sum, does not represent the true intensity and direction of the original magnetization of the rock. VRM is proportional to the square of the logarithm of the time of exposure to an external field and it also decays with time at the same rate as it is acquired when the field is removed. The viscous magnetization that arises in some rocks from exposure to the earth's field in the course of several months may be 100 times the induced magnetization, 10–15% of the NRM and 4–15% of TRM produced in the laboratory. Viscous magnetization is inversely related to coercive force. For example, magnetite ores which are magnetically softer than basalt develop in 70 days a VRM as large as 18–20% of the TRM (Sholpo, 1967).

NATURAL MAGNETIZATION OF OCEANIC BASALTS

Magnetic properties of oceanic basalts have been measured on samples dredged from either seamounts or submarine escarpments. Some more values may be expected from the deep-sea drilling program. These specimens are often fine-grained and glassy from rapid cooling in sea water which endows them with a lower magnetic susceptibility and a higher coercive force than similar continental basalts (Cox and Doell, 1962; Ade-Hall, 1964; Ozima et al., 1968). The mean magnitude of their NRM agrees well with the NRM calculated from seamount anomalies by the method of Chapter 7 as well as with that of continental basalts. In Appendix 3 presenting these data, Q is the "Koenigsberger ratio" which is the ratio of the NRM divided by the magnetization induced by the earth's field taken as 0.5 Oe; i.e., $Q = J_n/0.5\ k$. As previously mentioned, k is small for submarine basalts making Q large compared to continental rocks. Fig.22 shows the distribution of a

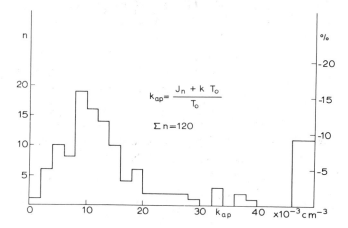

Fig.22. Apparent susceptibility of 120 specimens of submarine basalts. J_n is the natural remanent magnetization; k is the magnetic susceptibility; T_0 the value of the earth's magnetic field. (J. Francheteau, personal communication, 1970.)

new magnetic quantity, the "apparent susceptibility" which is useful for model calculations. The thermoremanent magnetization, being proportional to the external field can be thought of as the product of the geomagnetic field and an "effective" or "apparent" susceptibility.

Underwater photographs of submarine basalts extruded from ridges show that they often occur in pillow form. Exposures of pillow basalts about 1 km thick in Cyprus and Papua have been interpreted as uplifted sections of ocean floor, which would be more than adequate sources of lineated magnetic anomalies. Magnetic properties vary greatly with distance from the surface of the pillow (Marshall and Cox, 1971). The NRM of the outer 2 cm of a pillow 20 cm in average diameter is often only one fifth of that of the inner material. The glassy crust is naturally less magnetic and the magnetic minerals are concentrated toward the center of the pillow as it solidifies. To obtain a valid average value for the NRM from measurements on small specimens, one must make sure they represent a fair distribution with respect to depth in the pillow. In the case of pillows larger than 20 cm in diameter the NRM after rising to a maximum value decreases toward the center ostensibly because of increased crystal size arising from slower cooling. The average NRM of pillow basalts in middle latitudes measured by Marshall and Cox (1971) is between 0.014 and 0.007 e.m.u., and should be less for slowly-cooling massive dikes of same composition.

Irving et al. (1970) studied the magnetic properties of 38 fresh specimens of basalt collected across the Mid-Atlantic Ridge at latitude 45°N. They found an average NRM of 0.0092 c.g.s. units but the nine specimens from the central valley gave 0.0574 c.g.s. They suggest that partial oxidation of the titanomagnetite which can be accelerated by heating in· the laboratory occurs after extrusion of the pillow lavas with a resulting increase in the Curie temperature and a decrease of the NRM most of which takes place in the first 5 km from the axis of the central valley, but which can be felt as far as 150 km away. Irving et al. (1970) had the advantage of using a rock core drill rather than a dredge for collecting some of their specimens, which assured them of getting fresh rock.

Fortunately, interpretation of world-wide magnetic anomaly lineations at sea is not beclouded by uncertainties regarding the stability of rock magnetism or by self-reversal of TRM exhibited by a few rocks containing titaniferrous magnetite (Nagata, 1961, p.176; Uyeda, 1962). From observing the identity of the lineated anomaly pattern parallel to ridges in different oceans, we have sufficient proof that oceanic basalts have retained the record of geomagnetic field reversals for close to 80 m.y. A soft magnetization acquired since the last reversal of the geomagnetic field adds equally to the normally and to the reversely magnetized bands of rock and is thus undetectable from the surface of the sea. However, in the case of seamounts, VRM may take the direction of measured total magnetization appreciably different from that of the TRM the seamount acquired in the remote past.

INSTRUMENTS AND METHODS FOR MEASURING MAGNETIC PROPERTIES

The magnetic properties of oceanic rock samples are measured by the methods used on land rocks for paleomagnetic investigations. This subject is thoroughly reviewed in a collection of papers edited by Collinson et al. (1967).

The specimen to be measured for remanent magnetization is shaped into a cube or a circular cylinder of length roughly equal to its diameter. It is first measured as is and then again after magnetic "cleaning" for removing the viscous magnetism. This can be done by tumbling the sample in a steadily decreasing alternating magnetic field from some maximum value (Collinson et al., 1967, p.241). The procedure is repeated using increasing values of the maximum field. If it is a hard rock, cleaning can also be done by heating the specimen. In both methods, the ambient field is cancelled during the cleaning procedures by a system of coils.

The remanent magnetization is measured either with an astatic magnetometer or by spinning the specimen in a pickup coil and measuring the generated e.m.f.

The specimen is usually placed under the lower magnet of the astatic magnetometer (Fig.23A) and rotated about the vertical by remote control to four positions 90° apart.

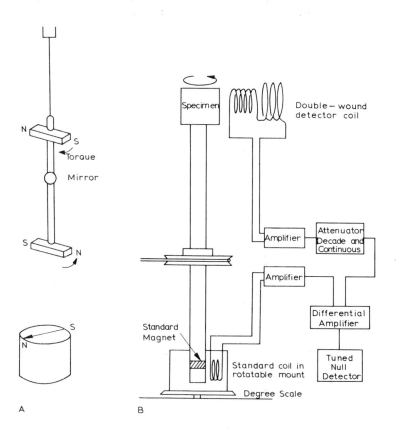

Fig.23. The measurement of magnetic remanence in rocks. A. The astatic magnetometer. B. A spinner-magnetometer. A specimen must be placed in various orientations in either instrument to determine its moment vectorially. (After Stacey, 1969.)

The measurements are repeated after inverting the specimen top to bottom. A similar set of eight readings is taken after turning the sample 90° about a horizontal axis. The three components of magnetization are calculated from the sixteen measurements (Collinson et al., 1967, p.172).

In the case of the spinner magnetometer (Fig.23B) or rock generator as it is sometimes called, the component of magnetization normal to the spin axis is measured for each position of the sample in the holder.

Two new rock-magnetism magnetometers were offered since the publication of Collinson et al. (1967). An improved spinner magnetometer capable of routine measurements down to 10^{-6} e.m.u. and perhaps a little better made by Princeton Applied Research Corp., Princeton, New Jersey has had much use. More recently, superconducting magnetometers (Develco, Inc., Mountain View, Calif.) have attracted much attention. A sensing coil in liquid helium in a magnetic shield picks up the magnetic flux due to the introduction of the sample. A sensitivity of $5 \cdot 10^{-9}$ e.m.u. is claimed for the superconducting magnetometer.

CHAPTER 3

MEASUREMENT OF GEOMAGNETIC INTENSITY AT SEA

THE TOTAL INTENSITY ANOMALY

In a moving body towed by a ship or an airplane it is impractical to measure components of the geomagnetic field with sufficient precision because the north and the vertical directions are difficult to establish. However, the magnetic anomalies seldom exceed 2% of the total earth's field, so that one can assume that the total magnetic vector stays undeflected over an area that is large in comparison to the wavelength of the anomalies, and that therefore its changes in magnitude can be treated as a vector quantity. Exactly the same kind of approximation is made in the case of gravity which is assumed to remain undeflected by the nonuniform lateral distribution of mass in the earth's crust. In Fig.24 let a geologic body produce an anomalous field ΔT which can be

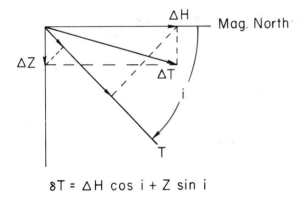

$$\delta T = \Delta H \cos i + Z \sin i$$

Fig.24. Approximate treatment of the total magnetic anomaly as a vector quantity. The anomalous magnetic intensity ΔT has components ΔH and ΔZ. The sum of their projections on the direction T of the geomagnetic field, δT is regarded as a vector in that direction, which is assumed to be the same over the area of the anomalies that are being interpreted.

split into a horizontal component ΔH and a vertical component ΔZ. What is actually measured is the difference between the planetary field T which in practice is the International Geomagnetic Reference Field (I.G.R.F.) and $|T + \Delta T|$. Most often geomagnetic model calculations deal with δT which consists of the sum of the projections of ΔH and ΔZ on the constant direction of T. The maximum errors that can be incurred by this approximation are given for different latitudes in Table II.

TABLE II

Maximum error in δT from assumption that the anomaly does not deflect the direction of the total magnetic intensity (units are γ)

(From Kontis and Young, 1964)

Total field \ Anomaly	500	1,000	2.000	3,000
30,000	4	17	67	150
35,000	4	14	57	129
40,000	3	13	50	113
45,000	3	11	44	100
50,000	3	10	40	90
55,000	2	9	36	82
60,000	2	8	33	75
65,000	2	8	31	69
70,000	2	7	29	64

THE PROTON MAGNETOMETER

Any total intensity magnetometer of sufficient precision can be used. A sensitivity of 5 or 10 γ is adequate for geomagnetic work at sea, since errors of the order of 30 γ might come from rapid time variations of the geomagnetic field and uncertainty of the ship's position. The modern proton precession magnetometers which are now universally used at sea record the field to one gamma at least once every minute. Except for the Russian nonmagnetic ship "Zaria" the hulls of oceanographic ships are made of steel, so that the magnetometer sensor has to be towed about three ship-lengths behind the ship to reduce errors on different headings to the 10-γ level (Bullard and Mason, 1961). The sensor consists of a coil immersed in a hydrogenous liquid such as a hydrocarbon oil or water. A d.c. current of several amperes, producing a field of over 100 Γ, is passed through the coil for several seconds to line up a small fraction of the protons, about one in 10^6, parallel to the axis of the coil. If this axis forms an angle to the direction of the geomagnetic field, when the d.c. field is shut off, the protons will start precessing about this field with a frequency proportional to its magnitude. The precessing protons induce an e.m.f. in the coil the frequency of which is measured by comparison with a crystal oscillator. The effect depends on the magnetic moment and angular momentum of the protons. The product of the magnetic moment of the proton and the component of the geomagnetic field perpendicular to it furnishes the torque which acts on the angular momentum to produce the precession. It corresponds to the gravitational torque acting on a pendulous gyroscope which is governed by the same equations of motion. It is easily demonstrated that the precession frequency is independent of the angle the gyroscope makes with gravity (Page, 1935, p.137). To make the amplitude less dependent on coil orientation, systems of several coils or a toroidal coil is used instead of a simple cylindrical coil which gives no signal when the coil axis is parallel to the geomagnetic field. The decrement of the amplitude of the precession signal differs for different liquids, being about 3 sec for water, and somewhat shorter for the oil mixtures. If the magnetic field T is measured in gammas, and the frequency in hertz:

$$T = 23.48682 \pm 8f \qquad \text{(N.B.S., 1971, chapter 2)}$$

The constant is proportional to the ratio of the angular momentum to the magnetic moment or the gyromagnetic ratio of the proton. One hertz corresponds to 23.5 γ so that if one wishes greater sensitivity, one has to measure fractions of a cycle. In the past this has been done by counting cycles of a high-frequency oscillator between a given number of precession cycles. The number of high-frequency cycles counted is proportional to the period of the precession, requiring conversion to field values from a table. Modern magnetometers multiply the precession frequency and count it in time units of such size that the counter reads directly in gammas. Recording proton magnetometers are now built by several manufacturers for station, ship, and aircraft use. A block diagram of such an instrument is shown in Fig.25, and a photograph in Fig.26. The only part of the proton precession magnetometer that might need checking is the oscillator in the counter. The oscillators in such counters are usually controlled by thermostated quartz crystals the

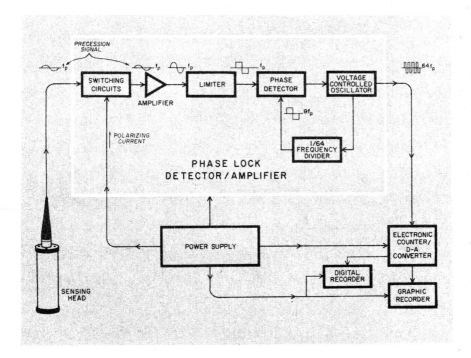

Fig.25. Block diagram of proton magnetometer Model V-4970. (From Varian Associates, Palo Alto, California, publication No. INS 1012A.)

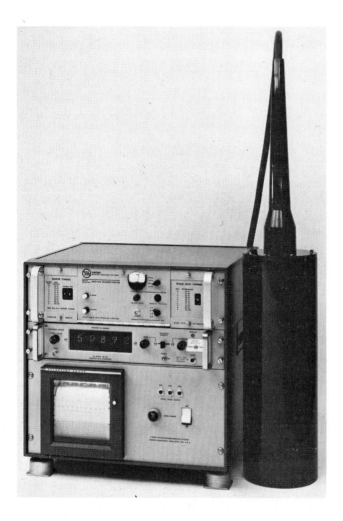

Fig.26. Photograph of Varian Associates, Palo Alto, California, shipboard proton magnetometer Model V-4970.

frequency of which remains constant to a few parts per million. It can be checked by a standard oscillator. The proton magnetometer has the great advantage over the oriented flux gate of not requiring calibration or measurement of base-line values. Its digital output is conveniently recorded on magnetic or punched paper tape, or directly fed into a shipboard computer system where the International Geomagnetic Reference Field can be subtracted from it to yield the anomalous field. Where the processing is not automatic, the magnetic record and the position on the track are correlated according to time after the ship's track is worked out from navigational data.

CHAPTER 4

THE EARTH'S MAGNETISM AND ITS HISTORY

THE ELEMENTS OF THE GEOMAGNETIC FIELD

At the general point on the earth's surface the geomagnetic field can be specified by a sufficient number of so-called elements depicted on Fig.27. From harmonic analysis of

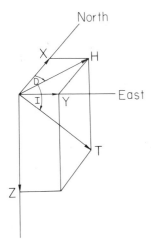

Fig.27. Conventional designation of geomagnetic elements. T = total intensity also called F; Z = vertical intensity; H = horizontal intensity; X = north component of H; Y = east component of H; D = declination; I = inclination or dip angle.

the few magnetic intensity measurements available to him, Gauss was able to prove in 1835 that most, if not all, of the geomagnetic field is of internal origin. Vestine et al. (1947) found that sources external to the earth contributed less than 1% to the global magnetic field. Despite the small size of the external field, it varies rapidly with time and thus limits the precision of marine surveys unless special recordings are made in the immediate vicinity of the survey to remove the time-varying part, as is done at times in the case of geophysical prospecting surveys on land, a procedure that is not economically practical at sea.

TIME FLUCTUATIONS

Rapid magnetic fluctuations are caused by electric currents in the ionosphere flowing at heights generally in excess of 100 km. The time-varying magnetic fields of these currents induce electric currents in the earth and in the water of the ocean which tend to oppose the changes of the magnetic fields of the sources. At the edge of islands and continents local magnetic fluctuations arise from the configuration of the coasts which affects the distribution of induced electric currents in the water making it difficult to use records of coastal magnetic observatories for correcting marine magnetic surveys, although this has been attempted for example in the Gulf of Aden (Whitmarsh and Jones, 1969).

There are two kinds of rapid magnetic fluctuations pertinent to marine surveys: the daily variation and the storm-type variation. The daily variation is produced by two large current whorls, one flowing in the upper, the other in the lower half of the sunlit hemisphere. They are produced by the motion of the ionosphere in the magnetic field of

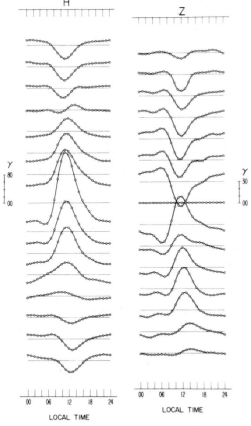

Fig.28. Average quiet day geomagnetic variations in H and Z during the equinox from 60°N to 60°S. The first curves from the geomagnetic equator are at approx. 5° latitude. After approx. 10° they are spaced 10° apart. (Matsushita, 1967, p.323.)

the earth. The motion is caused by tidal forces and expansion and ionization of the ionosphere by radiation from the sun. The average magnetic field variations are plotted against local time for different latitudes on Fig.28. We see that marine magnetic surveys may be subject to large errors in the auroral zone and in the equatorial belt in the vicinity of the electrojet, where errors as large as 200 γ can occur.

Magnetic storm variations happen simultaneously all over the earth. They are caused by a sudden increase in corpuscular emission from the sun. This corpuscular stream now called *solar plasma* contains as many positively charged particles as negatively charged ones. When it comes under the influence of the earth's magnetic field, the electrically charged particles become trapped and circulate in various ways forming the *magnetosphere*, thus causing erratic swings of the geomagnetic field reaching several hundred gamma amplitude and also some small quasi-periodic fluctuations called *micropulsations*. In addition to these phenomena the principal feature of a magnetic storm can be explained by the drift of the charged particles in the equatorial plane at a distance of six earth radii called the *ring current*. During the main phase of the storm the ring current is

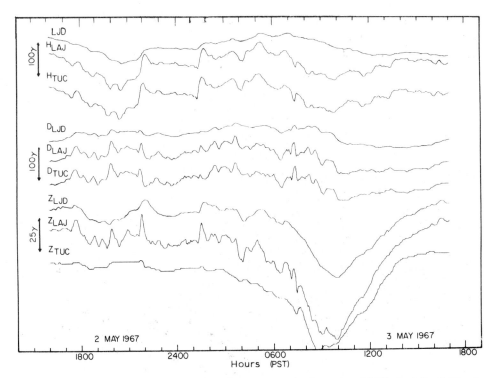

Fig.29. Record of a small magnetic storm at Tucson magnetic observatory (*TUC*), La Jolla, California (*LAJ*) and at the bottom of the ocean about 50 km seaward of the foot of the continental slope west of La Jolla (*LJD*). Note the attenuation of the high-frequency fluctuations at the deep water station (*LJD*) and their magnification in the Z component in *LAJ*. The magnification is attributed to "coast effect" (see Chapter 13) caused by electric currents induced in sea water and also perhaps due to probable seaward shallowing of isotherms in the earth's mantle. (Greenhouse, 1972.)

in such a direction as to oppose the dipole field. It takes several days for the ring current to decrease to its normal value. Thus, the mean value of the geomagnetic field is depressed during years of high sunspot activity which follows an 11-year cycle. Lately the structure and the behavior of the magnetosphere have been investigated by space probes (Matsushita, 1967, vol. II). The record of a small magnetic storm recorded at the bottom of the ocean is shown on Fig.29. The frequency of magnetic storms is governed by the 11-year sunspot cycle. On the average they might occur about 10 times per year. Days are characterized by magnetic activity measures one of which is the international index number k, which ranges from 0 to 9, and which is calculated at observatories from departures of the three magnetic elements from their three-hourly means. The values are adjusted according to the latitude of the observatory to make results from all observatories comparable. In compiling magnetic intensity data from marine surveys it is good practice to chart the k number as a measure of their reliability.

SECULAR VARIATIONS

Repeated world-wide observations of the geomagnetic field in historic times, as well as changes in its direction from the measurement of the magnetization of archeologically dated clays in pottery, pottery firing kilns, historic lava flows and varved clay deposits in

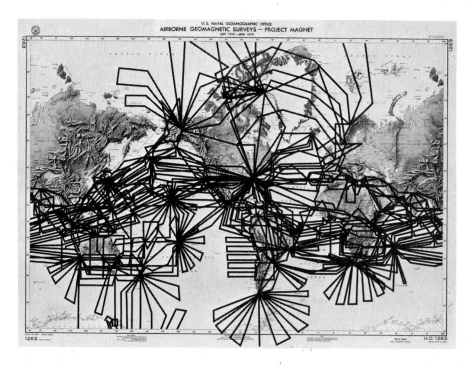

Fig.30. World coverage by Project Magnet as of April 1970 (Chart H. O. 1263, U.S. Naval Oceanographic Office).

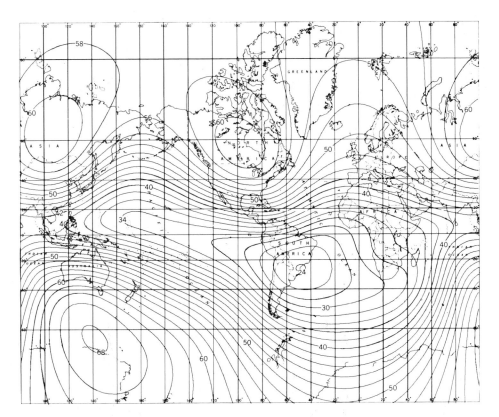

Fig.31. World chart of total magnetic intensity for 1965. Contours are labeled in tens milligauss. (Fabiano and Peddie, 1969.)

lakes, have shown that the geomagnetic field experiences secular changes in magnitude and direction. It is the custom to describe the earth's internal field as consisting of a dipole at the center of the earth directed toward the south geographic pole which produces a field of constant direction but slowly varying magnitude (see Fig.38) and of a smaller nondipole part which at any particular time is a function of position expressed in spherical harmonic series of latitude and longitude. The "variation" of the magnetic compass from geographic north is caused by this nondipole field. As we shall discuss it later on, there is unassailable experimental evidence that over 10,000 years, the nondipole part of the geomagnetic field is averaged out to zero by the secular variation leaving only the dipole part for magnetizing the large masses of rock which produce the magnetic field anomalies we measure. This means that departure of the thermoremanent magnetization of a large body of igneous rocks from the present direction of the dipole part of the geomagnetic field can be interpreted as caused by geographic displacement in latitude and azimuth of the site after the rock cooled down. It is the fundamental concept for interpreting paleomagnetism in terms of plate tectonics.

However, if we have the task of combining in the same area observations taken only a

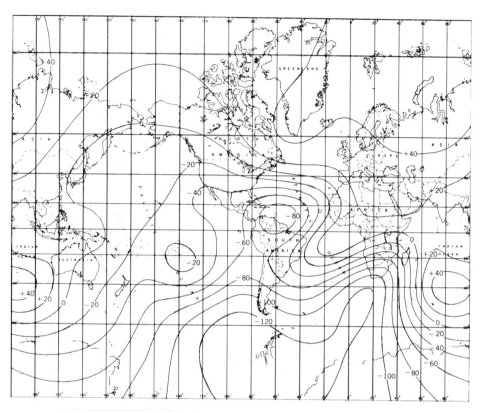

Fig.32. World chart of annual change of total magnetic intensity for 1965. The contour interval is 20 γ. (Fabiano and Peddie, 1969.)

few years apart, then secular variation which can be as large as 120 γ per year, has to be taken into account.

Secular variation is sufficiently rapid to require constant measurement of the magnetic field at magnetic observatories, repeat stations and by Project Magnet airplanes of the United States Navy Oceanographic Office. Information from these sources are combined for the publication of world charts of the magnetic elements every 10 years by the U.S. Coast and Geodetic Survey and by the British Admiralty. Fig.30 gives the world coverage of Project Magnet, the source of the most numerous magnetic data in the oceans. In the future, it is expected that the geomagnetic field and its secular variation will be obtained more accurately from total intensity measurements aboard low orbiting (1,500 km altitude) satellites. An uncertainty of 50 γ, 30 γ of which is attributed to the magneto-sphere has been calculated from existing satellite observations (Cain, 1971). The satellite observations cover the earth's surface uniformly and are undisturbed by anomalies from rock magnetism which often are 200–400 γ in amplitude. Fig.31 and Fig.32 show world charts of the total magnetic intensity and its annual variation for 1965. In the past, the reference field was obtained by interpolation from these charts. Nowadays there has been set up by international agreement the International Geomagnetic Reference Field, for which the harmonic coefficients are given in Appendix 1. The coefficients with their secular variation are inserted in appropriate computer programs to give the reference field for a given date at corners of a latitude-longitude grid. The field between the grid points is obtained by linear interpolation. Bullard (1967) shows that with a 1° grid interval the maximum error from the linear interpolation is 2 γ. The reference field so obtained is subtracted from the observed field in the computer to give the anomalous field.

CHRONOLOGY OF GEOMAGNETIC FIELD REVERSALS

The direction of the geomagnetic field in the geologic past is recorded in rocks and sediments in the oceans as well as on land. For several decades it has been known that the

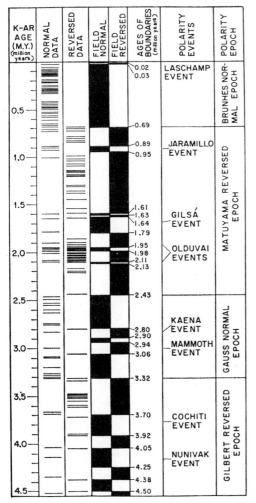

Fig.33. Chronology of geomagnetic field reversals from K-Ar age determinations on lava flows. (Cox, 1969.)

magnetization of successive lava flows on land was often reversed to the direction of the geomagnetic field, the normal and the reversed directions occurring in a definite pattern. These reversals were attributed to the geomagnetic field having a reverse orientation at the time the lavas cooled down, or to mineralogical mechanisms which produced a magnetization in the opposite direction to the ambient field during cooling. Intermediate directions were absent. Reverse magnetization from mineralogical mechanisms are now generally regarded as being rare isolated occurrences. By dating the lava flows by the potassium—argon method, one can establish the chronology of geomagnetic reversals back to about 4.5 million years before present (m.y.B.P.) before the errors of age determination become large enough to mask the reversals. Fig.33 shows the chronology of geomagnetic reversals so obtained. "Epochs" of normal and reversed polarity are interrupted at random by short "events" of opposite sign. The "epochs" and "events" have been given proper names which are now generally accepted in paleomagnetic literature. Some of the events are so short that their identification is subject to heated controversies (Watkins, 1972).

THE THEORY OF VINE AND MATTHEWS

Vine and Matthews (1963) and independently Morley and Larochelle (1964), proposed that the lineated magnetic anomalies found in the Northeast Pacific by Mason and Raff (1961) (Fig.34) were generated by the spreading of the sea floor postulated by Hess (1962) and Dietz (1962). Prior to this, Vacquier et al. (1961) found displacements of 1,400 km of the anomaly patterns across the Mendocino and Pioneer fracture zones which suggested mobility of rigid pieces of ocean floor, even though their original interpretation as transcurrent fault displacements was incorrect. The Vine and Matthews (1963) concept is illustrated on Fig.35 which shows a magnetic profile normal to the Juan de Fuca Ridge. Basalt rises at the ridge crest moving symmetrically outward, cools in the geomagnetic field acquiring a thermoremanent magnetization in the same direction as the field. When the field reverses direction, a band of oppositely magnetized basalt is laid down. If the spreading is uniform, geomagnetic reversals are recorded in the ocean bottom in a similar way to the tape of a magnetic tape recorder. The observed symmetry of the magnetic anomalies about the ridge axis supports the validity of the simple symmetrical spreading of Fig.35 and it is also found that the same idealized pattern of alternately magnetized strips produces computed magnetic anomalies similar to those observed in different parts of the world. The shape of the anomalies depends on the latitude and the orientation of the ridge, but the pattern of sea-floor magnetization reversals is the same everywhere with respect to time. This has been demonstrated beyond the shadow of a doubt (to most geophysicists at least) that alternate directions of magnetization of rocks and sediments have been caused by world-wide reversals of the geomagnetic field. Descriptions of lineated magnetic anomalies will be presented in detail in Chapters 5, 8—12. At this time we shall bring up only evidence needed to establish the

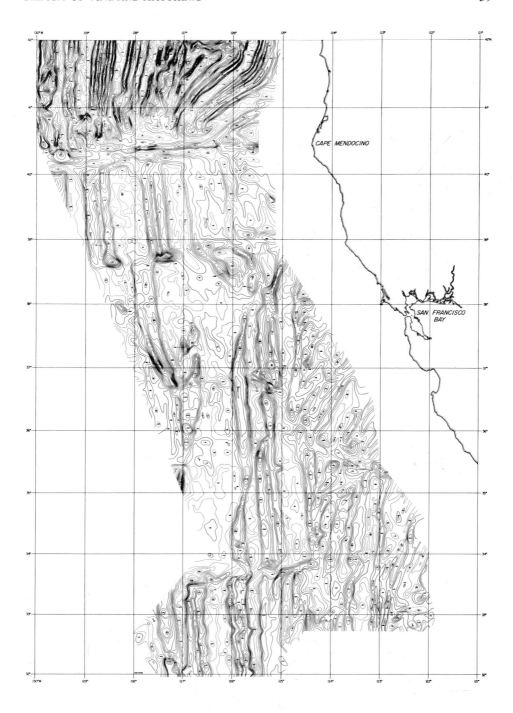

Fig.34. Total magnetic intensity anomaly chart off the west coast of California. Contour interval 40 λ. (Mason and Raff, 1961.)

character and the chronology of geomagnetic field reversals from data taken in the oceans.

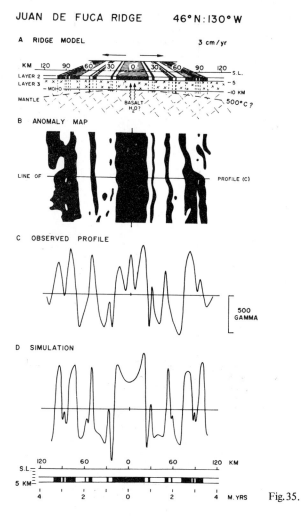

A. A schematic representation of the crustal model discussed in the text, applied to the Juan de Fucca Ridge, southwest of Vancouver Island. Shaded material in layer 2, normally magnetized; unshaded, reversely magnetized.

B. Part of the summary map of magnetic anomalies recorded over the Juan de Fuca Ridge (Raff and Mason, 1961). Black, areas of positive anomalies; white, areas of negative anomalies.

C. A total-field magnetic anomaly profile along the line indicated in B.

D. A computed profile assuming the model and reversal time scale discussed in the text. Intensity and dip of the earth's magnetic field taken as 54,000 γ and +66°; magnetic bearing of profile 087°. (1 γ = 10^{-5} Oe. $S.L.$ = sea level.)

Note: Throughout, observed and computed profiles have been drawn in the same proportion: 10 km horizontally is equivalent to 100 γ vertically. Normal or reverse magnetization is with respect to an axial dipole vector, and the effective susceptibility assumed is ±0.01 except for the central block at a ridge crest (+0.02). (From Vine, 1968.)

REVERSAL CHRONOLOGY IN OCEAN SEDIMENTS

The inclination of the direction of magnetization to the axis of a core of oceanic sediment can be measured by ordinary paleomagnetic methods. Some of the fine particles $1-10\,\mu$ in diameter constituting marine sediments come from weathering of rocks containing magnetic minerals, mostly titaniferous magnetite. Despite compaction and the thorough mixing of the first few cm of sediment by burrowing marine animals, the ancient geomagnetic field direction is faithfully recorded in cores of marine sediments. On the other hand, laboratory re-disposition in quiet water of the very same sediment usually leads to flattening of the magnetic inclination of the order of $5-20°$ because elongated particles are more magnetic than spherical ones and tend to settle with their long dimension horizontal. Even if the particles are spherical, if they roll some as they settle on the bottom, the average dip angle will be shallower than the inclination of the field (Griffiths et al., 1962). It is thus likely that marine sediments acquire their magnetization not during settling but after deposition. Experiments by Lloyd (quoted in Clegg et al., 1954) tend to confirm this hypothesis. Triassic marl was crushed to powder and allowed to settle in a tank. Magnetization of the resulting sediment showed a flattening of $8°$ in a field inclined $65°$. Sediment of the same marl deposited in zero field could also be magnetized by standing in a weak field. The magnetization of that sediment showed no inclination error provided the water content of the deposit exceeded 50%. A similar experiment is quoted by Irving (1964, p.32) on a mixture of magnetite and quartz sand grains ranging in particle size from coarse silt to fine sand. The ingredients were mixed dry, then flooded with water and allowed to stand over night in a weak magnetic field. The magnetization of the resulting sediment was parallel to the external field which had an inclination of $65°$. In all probability the depositional magnetization of marine sediments has been erased by subsequent reworking by animals, and the magnetism we measure has been acquired by standing in the earth's field while they still contained much water by a mechanism that is still obscure. The rate of deposition in the deep oceans away from continents is about a millimeter per thousand years, so that we may expect the secular variation to be nearly averaged out in a sample about 1 cm big. The rate of sedimentation determines the length of the time interval recorded in a deep-sea core. When sedimentation is slow as in Fig.36, most of the time scale of Fig.33 appears in the magnetization reversals of the core. The length in cm of polarity epochs in the core differs somewhat from the K–Ar scale, reflecting variations in rates of deposition. Nevertheless the pattern of field reversal epochs determined on continental lavas is evident in the sedimentary record in the ocean.

When the rate of sedimentation is rapid, it may reveal very short polarity events which are likely to be missed in a series of discrete continental lava flows. Thus reversal events have been found as short as 4,000 years. Rapid sedimentation also shows that the geomagnetic field switches sign in just a few thousand years as shown on Fig.37. It does so not by slowly swinging around, but by decreasing in intensity to nearly zero and then growing in the opposite direction. The intensity of the earth's field in the geologic past

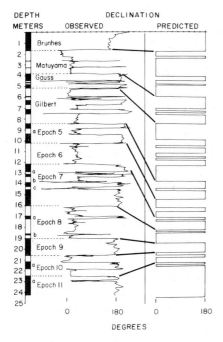

Fig.36. The change of magnetic declination in Conrad 12-65 after partial demagnetization of 50 Oe with respect to the split face of the core as a function of depth, compared with the reversal sequence predicted by the sea-floor-spreading time scale of Vine (1968). The black (normal) and white (reversed) bar diagram on the left indicates the proposed extension of the geomagnetic time scale based on this study. (Foster and Opdyke, 1970.)

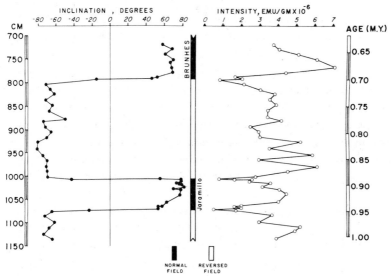

Fig.37. Character of the geomagnetic reversal process as recorded in a deep-sea sediment core. (After Opdyke, 1968.)

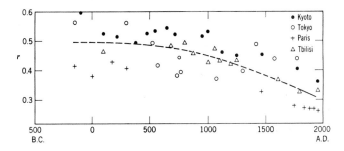

Fig.38. Intensity variation of the geomagnetic field over 2,000 years from measurement of four archeomagnetic groups. All values have been adjusted to give field intensities at the equator, assuming that the field was a simple dipole, a procedure which is expected to introduce up to 20% errors in individual observations but cannot affect the trend apparent in all of the data taken together. Figure from Sasajima (1965, quoted in Stacey, 1969) by permission of the Society of Terrestrial Magnetism and Electricity, Japan, and the author. (From Stacey, 1969.)

has also been investigated by measurements on rocks baked by geologically dated lava flows. In the study of Bolshakov and Solodovnikov (1969), the values of the ancient field for eight dated lava flows ranging from Late Quaternary to Oligocene vary from 0.48 to 1.05 of the present day value, whereas archeomagnetic results on ancient pottery and bricks suggest that about 200 A.D. the geomagnetic field was 1.6 times stronger than it is now (Burlatskaia et al., 1969). The paleointensity measurement consists of comparing the NRM of a specimen to the TRM it acquires in laboratory experiments from a weak field of known strength (Théllier and Théllier, 1959). Archeomagnetic measurements summarized in Fig.38 indicate the intensity of the geomagnetic field is now decreasing at a rate that will bring it to zero in about 2,000 years.

World-wide sampling of paleomagnetic directions of recent rocks has shown that on the average the geomagnetic field did not depart from the field of an axial dipole directed to geographic south in the last 10 m.y. Therefore if the magnetic inclination is faithfully recorded in oceanic sediments, it should be related to geographic latitude λ by $\tan I = 2 \tan \lambda$. That this is indeed true is proved by Fig.39 where the mean inclination from samples from the Brunhes and Matuyama epochs taken from 52 cores of ocean floor sediment obtained at various geographic latitudes are plotted with the inclination curve given by the formula.

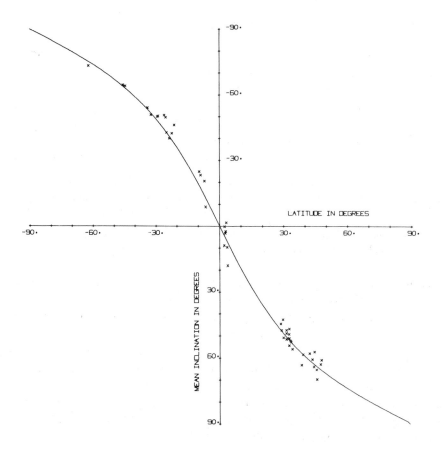

Fig.39. Magnetic inclination in 52 ocean cores for Brunhes and Matuyama epochs plotted against geographic latitude. The smooth curve is the theoretical relation tan I = 2 tan λ for the geocentric dipole. (From Opdyke and Henry, 1969.)

EXTRAPOLATION OF REVERSAL CHRONOLOGY FROM LINEATED ANOMALIES

So far we have built a time scale of geomagnetic field reversals based on K–Ar measurements and paleomagnetic measurements on lavas and oceanic sediments which extends to 4.5 m.y.B.P. To apply this scale to the Vine and Matthews (1963) method of measuring sea-floor spreading, we have to compare the anomaly computed from a model with the lineated anomaly pattern measured in the ocean. The model consists of parallel strips of normally and reversely magnetized rock parallel to the ridge crest, the top of which lies at the surface of the ocean bottom. The vertical thickness of the magnetized crust is usually taken as being constant and lying within the "basaltic" layer (Layer 2 of seismology), although it is conceivable that it could extend down to the Curie point isotherm, the maximum value of which is about 575°C (Vine, 1968). As a rule, the thickness of the model strip is taken between 0.5 and 2.5 km but since the ocean is generally 4 km deep, the shapes of the calculated anomalies are not seriously affected by the assumed thickness of the model. Recent measurements by Irving et al. (1970) on basalt samples from the Mid-Atlantic Ridge suggest a thickness of only 200 m for the magnetic layer. The models usually assume that the rock was magnetized by the field of the geocentric dipole at present latitude. This is not necessarily the case when north—south spreading of ancient age has displaced the site in latitude. The shape of the anomalies is also dependent on the azimuth of the profile. In Appendix 2, computation of sea-floor spreading magnetic anomalies from strip models is described in detail.

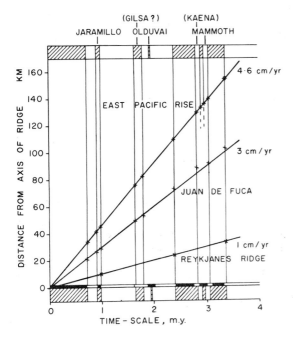

Fig.40. Inferred normal-reverse boundaries in the oceanic crust plotted against a suggested reversal time scale. (Shaded intervals, periods of normal polarity.) The thin black-and-white time scale indicated immediately above the abscissa is that suggested by Cox (1969). (From Vine, 1968.)

On Fig.40 the boundaries between the normal and reversely magnetized strips of the model are plotted against distance from the ridge crest for three different ridges. Along the horizontal axis the time scale of geomagnetic field reversals is also shown up to 4 m.y.B.P. The spreading rates given by the slopes of the lines appear to be uniform for the three ridges.

Heirtzler et al. (1968) assumed that the spreading rate in the South Atlantic remained constant for the whole sequence of recognizable magnetic anomalies. On this assumption, and from paleontological evidence of a South Atlantic core, the oldest recognizable magnetic anomaly was laid down about 80 m.y. ago. The magnetic anomalies which were

TABLE III[1]
Intervals of normal polarity (m.y.)
(After Heitzler et al., 1968)

0.00- 0.69	18.91-19.26	40.03-40.25
0.89- 0.94	19.62-19.96	40.71-40.97
1.78- 1.93	20.19-21.31	41.15-41.46
2.48- 2.93	21.65-21.91	41.52-41.96
3.06- 3.37	22.17-22.64	42.28-43.26
4.04- 4.22	22.90-23.08	43.34-43.56
4.35- 4.53	23.29-23.40	43.64-44.01
4.66- 4.77	23.63-24.07	44.21-44.69
4.81- 5.01	24.41-24.59	44.77-45.24
5.61- 5.88	24.82-24.97	45.32-45.79
5.96- 6.24	25.25-25.43	46.76-47.26
6.57- 6.70	26.86-26.98	47.91-49.58
6.91- 7.00	27.05-27.37	52.41-54.16
7.07- 7.46	27.83-28.03	55.92-56.66
7.51- 7.55	28.35-28.44	58.04-58.94
7.91- 8.28	28.52-29.33	59.43-59.69
8.37- 8.51	29.78-30.42	60.01-60.53
8.79- 9.94	30.48-30.93	62.75-63.28
10.77-11.14	31.50-31.84	64.14-64.62
11.72-11.85	31.90-32.17	66.65-67.10
11.93-12.43	33.16-33.55	67.77-68.51
12.72-13.09	33.61-34.07	68.84-69.44
13.29-13.71	34.52-35.00	69.93-71.12
13.96-14.28	37.61-37.82	71.22-72.11
14.51-14.82	37.89-38.26	74.17-74.30
14.98-15.45	38.68-38.77	74.64-76.33
15.71-16.00	38.83-38.92	
16.03-16.41	39.03-39.11	
17.33-17.80	39.42-39.47	
17.83-18.02	39.77-40.00	

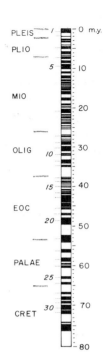

Fig.41. The geomagnetic time scale. From left to right; time scale for geologic eras, numbers assigned to bodies and magnetic anomalies, geomagnetic field polarity with normal polarity periods colored black. (Heirtzler et al., 1968.)

[1] McKenzie and Sclater (1971) identify a small positive anomaly after 32 and a strong positive 33 on most profiles in the North Pacific and Indian Oceans. Starting with the third line from the bottom of Table III, their modification of the Heirtzler et al. time scale is: 71.22–72.01; 74.01–74.21; 74.35–75.86; 76.06–76.11; 76.27–.

considered easy to identify were given reference numbers, as shown on Fig.41 and Table III, which have been adopted in the literature. Fig.42 presents the actual and the computed anomalies for the North and the South Pacific from presently spreading ridges. From the heavy vertical lines connecting the same anomalies on different profiles, it is apparent that variations of spreading rates have occurred. By comparing spreading rates in different oceans one might spot the places where changes of spreading rates have happened. This is done in Fig.43 where the distance to a given magnetic anomaly in the

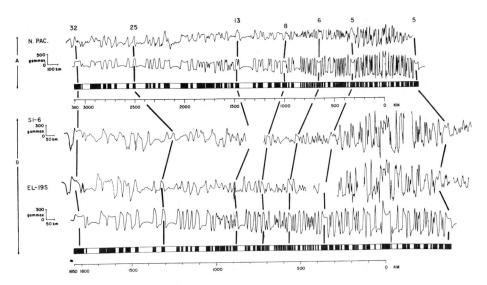

Fig.42. A. North Pacific composite profile. Bottom is the set of model blocks for the North Pacific, immediately above which is the computed anomaly profile. B. Top shows the *SI-6* and the *EL-19S* magnetic anomaly profiles from the South Pacific. Bottom is the set of model blocks derived for the South Pacific and immediately above is the computed magnetic profile. Time scale is related to distance by extrapolation from the Gilbert reversed epoch. (Pitman et al., 1968.)

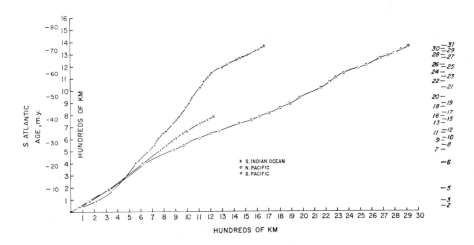

Fig.43. The distance to a given anomaly in the South Atlantic versus distance to the same anomaly in the South Indian, North Pacific, and South Pacific oceans. Numbers on right refer to anomaly numbers. (Heirtzler et al., 1968.)

South Indian, North Pacific and South Pacific Oceans is compared to the distance to the same anomaly in the South Atlantic. The three curves should show coherent departures from linearity if the spreading rate varied in the South Atlantic. After anomaly 6 (21 m.y.B.P.) spreading in the North Pacific appears to have slowed down by a ratio of about

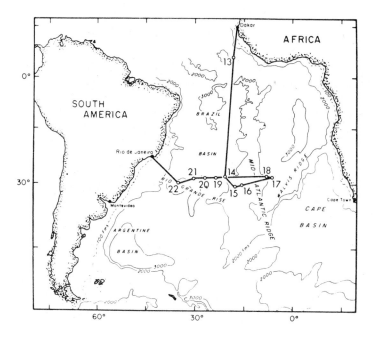

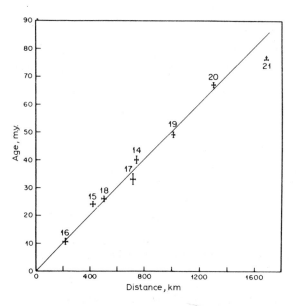

Fig.44. Comparison of paleontológical and magnetic anomaly ages from deep-sea drilling in the South Atlantic. The paleontological age of the sediment immediately above the basalt is plotted against distance from the ridge axis. The index number of the points refers to the drilling locations shown on the map. Slightly revised from Peterson (1969).

TABLE IV

Paleontological age of sediment from JOIDES cores immediately above basalt in the South Atlantic compared with extrapolated age of magnetic anomalies of Heirtzler et al. (1968)

Site no.[1]	Magnetic anomaly no.	Age from magnetic anomaly time scale (m.y.)	Paleontological age of sediment above basement (m.y.)	Distance from ridge axis along arc about rotation center at 62°N 36°W
16	5	9	11 ± 1	221 ± 20
15	6	21	24 ± 1	422 ± 20
18	–	–	26 ± 1	506 ± 20
17	–	34–38	33 ± 2	718 ± 20
14	13–14	38–39	40 ± 1.5	745 ± 10
19	21	53	49 ± 1	1010 ± 10
20	30	70–72	67 ± 1	1303 ± 10
21	–	–	76	1686 ± 10

[1] Basement rock not reached at site 21.

3/2 from 4.4 cm/year to 2.9 cm/year as pointed out by Vine (1966). The other two curves do not show a kink at this ordinate, so we can presume it happened in the North Pacific. Similarly we can argue that the spreading rate increased in the South Pacific about anomaly 24 time (60 m.y.B.P.). Tentatively one might conclude that spreading rates stay the same for long stretches of time. The geomagnetic reversal time scale of Heirtzler et al. (1968), as given on Fig.41 was confirmed by deep-sea drilling in the South Atlantic (Table IV, Fig.44). Since its publication in 1968 the first 10 m.y. of the Heirtzler

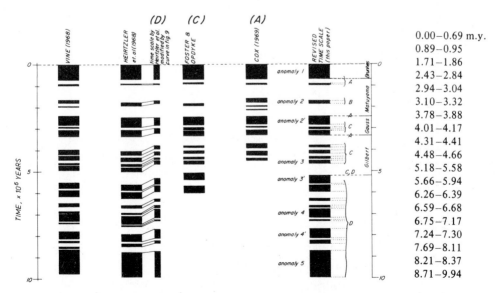

Fig.45. Proposed reversal chronologies. The revised chronology is shown at the right. The dotted lines to the right of this plotted time scale serve to indicate the source of the age for a particular reversal. Source B is Opdyke and Foster (1971, quoted in Talwani et al., 1971). The revised intervals of normal polarity are given in the table to the right. (Talwani et al., 1971.)

et al. geomagnetic reversal scale has suffered several revisions. Fig.45 reviews them and gives the latest revised intervals of normal polarity.

LENGTH OF POLARITY EPOCHS

Within the span of 80 m.y. covered by the reversal time scale, the geomagnetic field spent approximately the same amount of time in the normal and the reversed polarities (Fig.46). On the average, the magnetic field spends 0.45 m.y. in each polarity which

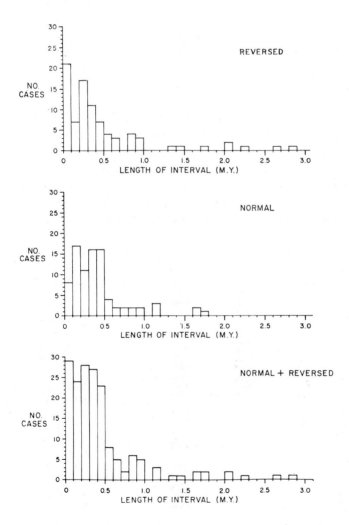

Fig.46. Distributions of the lengths of magnetic states for reversed polarity and normal polarity and for the lengths of intervals regardless of polarity. (Heirtzler et al., 1968.)

appear to be equally stable. The frequency of geomagnetic reversals, however, has shown a marked increase (Fig.47). The increase in frequency of oscillation of the anomalous field as one approaches the ridge also is evident on Fig. 42.

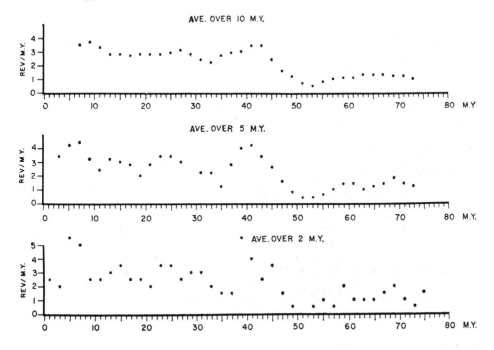

Fig.47. Frequency of geomagnetic field reversals as a function of time averaged over intervals of 2, 5, and 10 m.y. (Heirtzler et al., 1968.)

THE GEOMAGNETIC DYNAMO

Bullard (1968) discusses the impact of these data on the theories of the origin of the geomagnetic field. Convection of the electrically conducting substance of the earth's liquid core from an unspecified source of energy under the influence of the earth's rotation and in the presence of an infinitesimally small magnetic field produce electric currents which regeneratively grow to a steady value to give the present geomagnetic field (Elsasser, 1955). The mechanism is presumed to be analogous to the behavior of a disc homopolar generator of Fig.48, in which the current and magnetic field may point in either direction.

Two such dynamos, rotating at different speeds and connected in series as shown on Fig.49, can give by proper choice of circuit constants an oscillating electric current which will periodically change its sign as shown on Fig.50. The extension of this model to the spherical case of the earth's liquid core is very complex and has not as yet been carried out. Quoting Bullard (1968): "It is possible that the essential electromagnetic features of the mechanism producing the earth's field are simulated or realistically caricatured by the double dynamo".

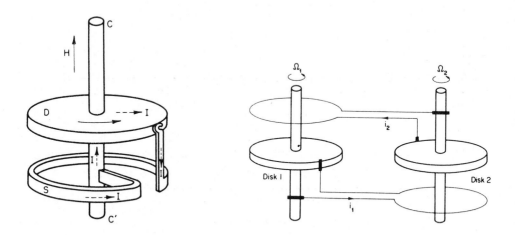

Fig.48. Bullard disk dynamo. (Bullard, 1968.) CC' = axis; D = rotating disk; S = spiral wire.

Fig.49. Two Bullard disk dynamos connected in series. (Rikitake, 1968.)

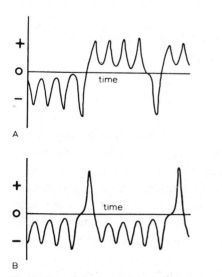

Fig.50. One of the currents in a pair of coupled disk dynamos. In (A) the current oscillates about one equilibrium value and then reverses and oscillates about the other. In (B), which is an earlier part of the same solution, the reversal only lasts for a short time. (From Allan, 1962, quoted in Bullard, 1968.)

CHAPTER 5

EXAMPLES OF LINEATED MAGNETIC ANOMALIES IN THE OCEANS

We shall examine in detail the more obvious evidence for sea-floor spreading contained in the pattern of lineated magnetic anomalies, first discovered by Mason and Raff (1961) (Fig.34) from which vast areas of the ocean floor can be dated using the chronology of geomagnetic field reversals discussed in the last chapter.

IDENTIFICATION OF LINEATED ANOMALIES

From the reversal chronology and a uniform spreading rate, a magnetic model of the ocean floor can be built consisting of normally and reversely magnetized rectangular strips of infinite length in the direction at right angles to the direction of spreading. The magnetic anomaly computed from the model is made to fit the observed profile by adjusting the spreading rate (Fig.42).

The latitude and the orientation of the model with respect to north influence the shape and magnitude of the computed anomalies. On Fig.51 the North and the South Atlantic profiles are nearly at the same latitude but as their azimuths are $150°$ and $116°$, respectively, the forms of the anomalies are substantially different. In equatorial latitudes some of the anomalies like the central one, reverse sign as shown by the bottom set of profiles. Furthermore, ship tracks are seldom straight and perpendicular to the strike of the magnetic anomalies. For these reasons it is generally inappropriate to compare directly observed magnetic profiles with each other or with the model anomaly as has often been done in published literature. To establish the direction of spreading, one needs at least two ship or airplane tracks forming a substantial angle with the strike of the anomalies. The strike is obtained by locating on the chart the same anomalies on the two tracks and joining them with lines which should be roughly parallel to each other if fracture zones are not crossed. The best direction of spreading which is perpendicular to the lines joining the anomalies is estimated and drawn on the chart. The magnetic curves on the tracks are projected on the line and compared to the model anomaly obtained from the standard geomagnetic field chronology for that particular latitude and azimuth. A uniform spreading rate that gives the best fit is tried first. Occasionally the spreading rate varies for different portions of the profile, as for example in the North Atlantic (see Fig.125). For direct comparison with the observed trace, the computed anomaly from the standard model can be projected onto the ship's track.

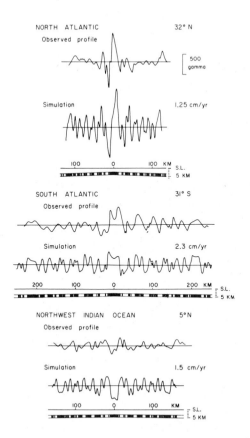

Fig.51. A comparison between observed and computed magnetic profiles at three widely separated points on the mid-ocean ridge system. Observed profiles from Heirtzler and Le Pichon (1965, quoted in Vine, 1968) and Matthews et al. (1965). The simulations assume the reversal time scale listed in Fig.41, a constant rate of spreading, and values for the intensity and dip of the earth's field and the magnetic bearing of each profile as follows: North Atlantic, 44,000 γ + 56°, 150°; South Atlantic, 27,600 γ − 54°, 116°; northwest Indian Ocean, 37,620 γ − 6°, **44° (Vine, 1968).**

SUBJECTIVITY OF THE COMPARISON PROCESS

Accurate calculations of standard anomalies help but do not remove the subjectivity of the process of comparing one curve with another one. Having found the appropriate portion of the model, one stretches one curve vertically and horizontally to fit another one until it "looks" right. Bullard (1968) wrote in this connection: "To predict the results of magnetic surveys from measurements on rocks thousands of miles away is a claim so exorbitant in a subject where things are rarely clear-cut or predictable, that it is necessary to be very sure that we are not deceiving ourselves. The agreement of the calculated and observed curves is so good that it seems impossible to ascribe it to chance. One could wish in this and other geophysical contexts that there were objective criteria for describing the resemblance of curves; the matter is one of some difficulty, the obvious criteria, using cross-correlation, are not very useful because they are dominated by the central peak and give little weight to the agreement of small peaks. Also cross-correlation is grossly disturbed by a little irregular stretching of the curves such as might be produced by slight variations in the velocity of spreading of the ocean floor. Similarity is not a simple idea, and is only partly metrical; it is concerned with the number and sequence of maxima, minima and zero crossings as well as with their positions and amplitudes. It may be quite difficult to formalize intuitive impressions about it, perhaps ideas from the detection of bomb tests by seismic arrays and from numerical taxonomy may be fruitful". And, indeed, experience shows that the subjective methods used to find similarities of magnetic profiles caused by sea-floor spreading need no justification when talking to seismologists.

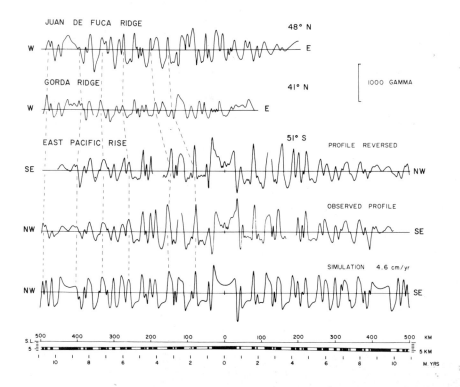

Fig.52. The "Eltanin"-19 profile across the East Pacific Rise (Pitman and Heirtzler, 1966) together with the profile reversed about its midpoint to demonstrate its symmetry, and a computed profile assuming the reversal time scale for the past 11 m.y. listed in Fig.41, a total field intensity and dip of 48,000 γ and $-62.6°$ respectively, and a magnetic bearing of 102° for the profile. The profile is also compared with a composite profile across and to the northwest of the Juan de Fuca Ridge, and a profile normal to the strike of the anomalies across and to the west of the Gorda Ridge (Raff and Mason, 1961; Vacquier et al., 1961). (From Vine, 1968.)

TEST FOR SYMMETRY

The Vine and Matthews (1963) method of measuring sea-floor spreading implies a symmetrical anomaly pattern on both sides of a spreading ridge, and a world-wide distribution of magnetic anomalies given by the same pattern of reversals of magnetization of the ocean bottom which differs from place to place only by the spreading rate usually taken as constant for each observed pattern. Symmetry is best demonstrated by plotting the same profile in the reverse direction and observing the correspondence of the individual features at the same distance from the central anomaly. This is done for the East Pacific Rise at 51°S in Fig.52. This profile can be compared to the ones plotted for the Gorda and the Juan de Fuca ridges in the northeast Pacific. Note that the amplitudes show the expected increase with latitude. The correlation of the three observed profiles and the computed profile is considered excellent.

The departures of the observed profile from the profile calculated from a model consisting of positively and negatively uniformly magnetized strips of infinite length and rectangular cross-section can arise from various causes like for example the intrusion of dikes of the opposite polarity (Harrison, 1968), as well as departures from idealized geometry assumed for the models the principal ones of which are sloping instead of vertical contacts, relief of the bottom surface and local topography. The effect of the latter is best demonstrated by profiles parallel to the strike located within a band of known age and same polarity. Where the sediment is thin like in the crestal area of a

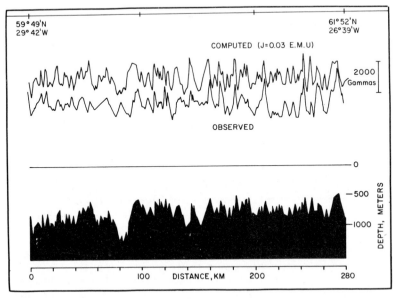

Fig.53. Depth and observed residual magnetic intensity profile along ridge crest. A computed profile, assuming that the anomaly is caused by two-dimensional uniformly magnetized topography with a remanent magnetization of $J = 0.03$ e.m.u., is shown at the top. (Talwani et al., 1971.)

spreading ridge, the outline of the surface of the basaltic layer can be gotten by seismic profiling, from which the magnetic profile at the surface of the ocean can be computed on the assumption that the depth normal to the profile remains constant. Despite this crude assumption, generally good agreement between observed and calculated profiles have been obtained on the Reykjanes Ridge (Fig.53).

THE ANTARCTIC PACIFIC

The coverage of the Antarctic Pacific as of 1968 is given in Fig.54, the heavier portions of the tracks indicating the segments used in plotting magnetic profiles across the ridge. The magnetic pattern resulting from sea-floor spreading is so persistent that it can be fairly well defined by the relatively few tracks of Fig.54, an impossible feat for a magnetic survey on land covering a comparable area. Consequently, we may expect that during the next decade, the world's oceans will be adequately surveyed for dating their floors and for locating the major breaks in the pattern caused by fracture zones. The Antarctic Pacific magnetic profiles, plotted to the same distance scale, are assembled in Fig.55 with the central anomaly aligned vertically at the center of the figure. In addition to the displacement of anomalies caused by fracture zones, the spreading rate steadily decreases from track EL 21 to track SI 5. This is also apparent on Fig.56 which is an enlargement of the central portion of Fig.55. As we shall see later, the phenomenon is

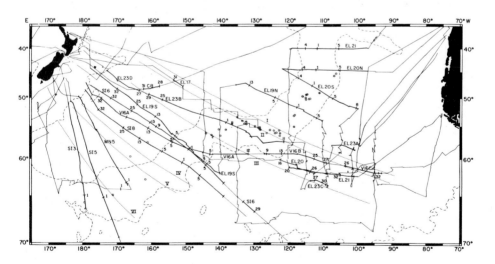

Fig.54. Ships' tracks in the antarctic region of the Pacific Ocean; the heavier parts of track refer to sections used for plotting magnetic profiles. The numbers refer to specific anomalies for track location. The 2,000-fathom contour is given by dashed lines; the dotted lines show the location of proposed fracture zones, each annotated with a roman numeral. The circles with the dots show the location of earthquake epicenters from June 1, 1954 to May 31, 1965. *EL* = "Eltanin"; *MN* = "Monsoon"; *SI* = "Staten Island"; *V* = "Vema"; *C* = "Conrad". (Pitman et al., 1968.)

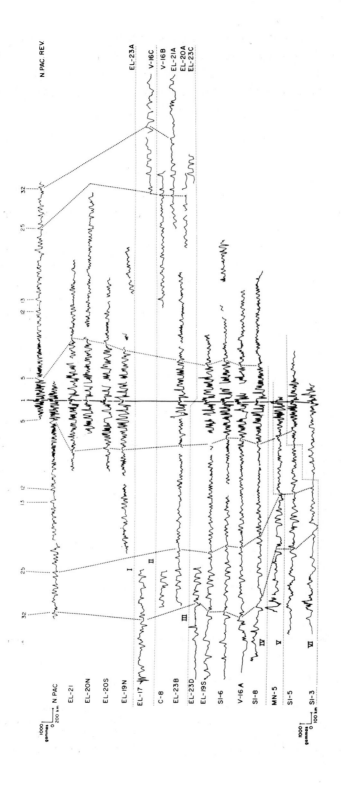

Fig.55. Observed profiles of magnetic anomalies along the tracks shown in Fig.54. Profiles have been projected along an azimuth normal to the ridge axis and aligned along the ridge axis denoted by the number *I* and the solid vertical line. The numbers correspond to the numbers in Fig.54. The topmost profile labeled *N. PAC.* and *N. PAC. REV.* (reversed) is from the North Pacific. The horizontal dotted lines with the roman numerals correspond to the fracture zones shown in Fig.54. (Pitman et al., 1968.)

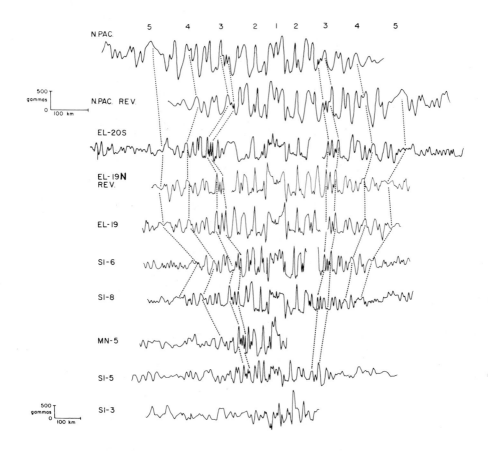

Fig.56. Observed magnetic anomaly profiles from Fig.55 along tracks shown in Fig.54 from the North Pacific. The numbers refer to specific anomalies; the dashed lines show the correlation. (Pitman et al., 1968.)

basic to the theory of plate tectonics because spreading rates are proportional to distance from the center of relative rotation between two lithospheric plates. The Antarctic Pacific is the only area where the complete set of sea-floor spreading magnetic anomalies up to anomaly 32 is present on both sides of center. To be able to recognize the similarity of observed magnetic anomalies to the anomalies computed from the standard model, one needs to plot them to a sufficiently open scale to bring out the individual character of the anomalies, as exemplified by Fig.57—59. Fig.59 shows anomalies beyond No.32 in time, posing an interesting question as to why correlation does not extend beyond that anomaly.

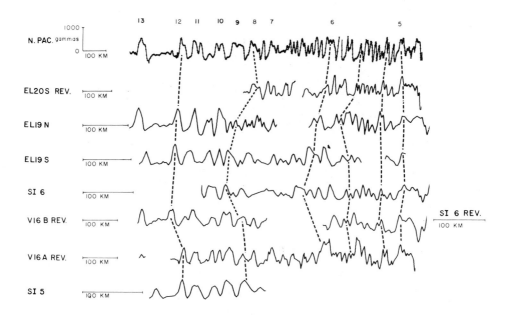

Fig.57. Observed anomaly profiles (anomalies 5–13) from Fig.55 along tracks shown in Fig.54 and from the North Pacific. Profiles from west of the ridge axis have been plotted with west to the left. Profiles from east of the ridge axis have been reversed (west on the right). The numbers refer to specific anomalies; the dashed lines show the correlation of some of these anomalies. (Pitman et al., 1968.)

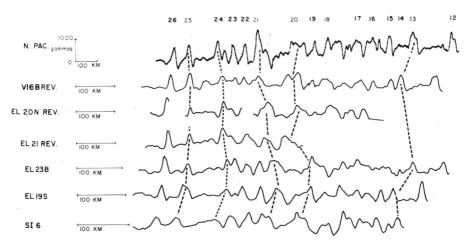

Fig.58. Observed anomaly profiles (anomalies 12–26) from Fig.55 along tracks shown in Fig.54 and from the North Pacific. Profiles from west of the ridge axis have been plotted with west to the left. Profiles from east of the ridge axis have been plotted with west to the right. The numbers refer to specific anomalies; the dashed lines show the correlation of some of anomalies. (Pitman et al., 1968.)

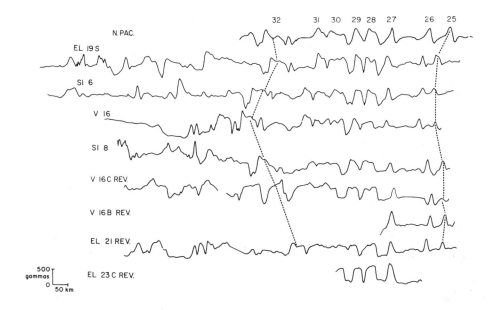

Fig.59. Observed anomaly profiles (anomalies 25–32) from Fig.55 along tracks shown in Fig.54 and from the North Pacific. Profiles from west of the ridge axis have been plotted with west to the left. Profiles from east of the ridge have been plotted with west to the right. The numbers refer to specific anomalies, the dashed lines show the correlation of some of these anomalies. (Pitman et al., 1968.)

THE SOUTH ATLANTIC

From the point of view of completeness of the sea-floor spreading record on both sides of the ridge, the South Atlantic comes next. Fig.60 presents ship tracks in that area along which the standard anomaly numbers are indicated, the letter A showing the position of the ridge. Again, one cannot help being impressed by the vast distances between cor-related profiles. The actual profiles west of the ridge and their correlation with the North Pacific lineations appear in Fig.61. The profiles east of the ridge are shown on Fig.62. A break in the anomaly sequence can be expected where a profile crosses a fracture zone; for example, anomaly 6 follows immediately anomaly 5 on profile V 22 on Fig.61. Detailed comparisons between profiles are made in Fig.61–64.

The interpretations discussed so far consisted of simply identifying anomalies from one profile to the next without recalculation of the model anomaly for each change of latitude and spreading direction. If enough profiles are available, the shapes of the anomalies change with position gradually, making it possible to identify them without resorting to a more elaborate method.

If the sea floor is spreading from north–south ridges as in the Northeast Pacific or the South Atlantic, the anomalies are observed at the latitude at which they originated. This is not true when spreading had a substantial north—south component and if furthermore, the piece of lithosphere carrying the anomaly had rotated about the vertical. Under these circumstances anomalies are difficult and sometimes impossible to recognize without computing several models for the most likely original positions and orientations. A complete description of a more elaborate method of comparison is given in Appendix 2.

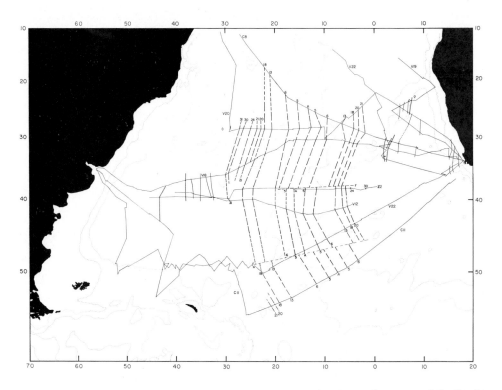

Fig.60. The location of magnetic anomalies along ship tracks for the southern part of the South Atlantic Ocean. *F* is the trace of a proposed fracture zone. The trace of linear anomalies in the Argentine basin and Cape basin are indicated by solid lines. Dotted lines are the 500- and 2,000-fathom isobaths. (Dickson et al., 1968.)

Fig.61. Magnetic profiles from the South Atlantic west of the Mid-Atlantic Ridge and a profile from the North Pacific. The ship tracks and numbered traces can be identified on Fig.60. Vertical scale in gammas. (Dickson et al., 1968.)

Fig.62. Magnetic profiles from the South Atlantic east of the Mid-Atlantic Ridge and a profile from the North Pacific. The ship tracks and numbered traces can be identified on Fig.60. Vertical scale in gammas. (Dickson et al., 1968.)

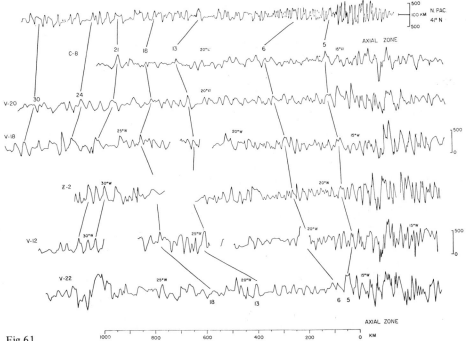

Fig.61.

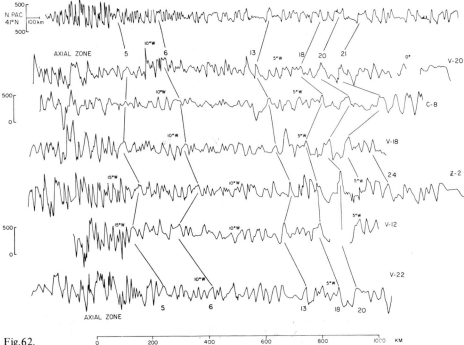

Fig.62.

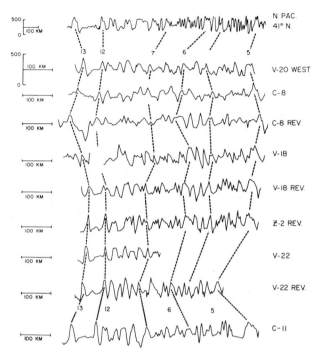

Fig.63. Detailed comparison of the anomalies between 5 and 13 for the South Atlantic Ocean and the North Pacific Ocean. Vertical scale in gammas. (Dickson et al., 1968.)

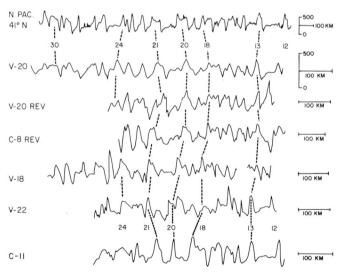

Fig.64. Detailed comparison of the anomalies from 13 to the end of the magnetic pattern in the South Atlantic Ocean and the North Pacific Ocean. Vertical scale in gammas. (Dickson et al., 1968.)

CHAPTER 6

UNDERTHRUSTING OF THE LITHOSPHERE AND PLATE TECTONICS

If the earth is to retain its surface area, as much area should be destroyed as is being created by sea-floor spreading, for otherwise the size of the earth would be growing at a preposterous rate. Excellent seismic evidence tells us that the lithosphere is absorbed by underthrusting mostly at deep oceanic trenches to compensate its widening at the ridges. "Subduction", as this consumption process is called, and spreading of the lithosphere are united in the theory of plate tectonics (McKenzie and Parker, 1967; Morgan, 1968; LePichon, 1968). As soon as it was demonstrated (Vacquier et al., 1961) that the same magnetic anomaly pattern was recognizable over distances of the order of 1,000 km on both sides of a fracture zone, it became evident that the ocean floor was not a viscous medium in which continents could float about, but that it was just as rigid as the continents. In plate tectonics, spreading at the East Pacific Rise, for example, is accommodated in part by the Aleutian and western Pacific trenches clear across the ocean.

Speculations on the mechanism responsible for the plate motions are not within the scope of this book. It is generally agreed that lateral density variations causing convective flow of mantle material arise from lateral variation of temperature. In addition to thermal expansion, density changes occur from phase changes. Radioactivity is an ample source of energy but its distribution with depth is not accurately known, except that elementary calculations based on measurements of radioactivity of rock specimens prove that the radioactive elements have been concentrated toward the surface by selective crystallization from rock melts. In the last ten years, estimates of the viscosity of the mantle as a function of depth have been steadily revised downward until mantle-wide convection, that is, convection below 400 km, is being talked about (Carter and Ave'Lallemant, 1970). One or two indeterminable per cent of water allows one to change the rheological properties of mantle rocks enough to fit everybody's calculations on mantle flow. This freedom in turn affects thermal calculations because in the presence of flow, the laboratory-measured thermal conductivity of solid rocks constitutes a minor part of the heat-transfer, the major part coming from the movement of the material. In view of the inherent uncertainties regarding the value of the physical constants in convection calculations, the latter do not support or disprove the factual magnetic, seismic, volcanological, gravitational and physiographic evidence for mantle flow.[1]

---

[1] The interested reader may consult the special issue of the *Journal of Geophysical Research*, 76, No. 5, 1971 on *Mechanical Properties and Processes of the Upper Mantle*.

MODEL OF SPREADING AND CONSUMPTION OF LITHOSPHERE

Sea-floor spreading and destruction are illustrated diagrammatically in Fig.65 in

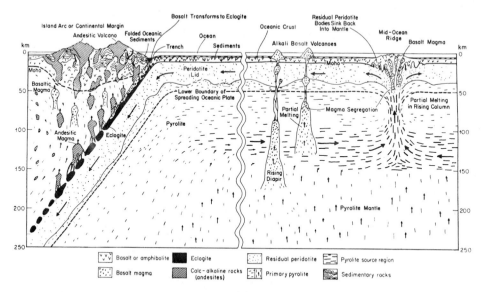

Fig.65. Diagrammatic sketch of sea-floor spreading and absorption hypothesis. The depth of the return flow shown by the arrows in the middle of the figure is unknown (see text). (After Ringwood, 1969.)

petrological terms. The motion of partially molten mantle rock drags the rigid lithospheric plate, consisting of a thin layer of basalt on a thicker plate of peridotite, away from the ridge crest. The plate sinks into the pyrolite mantle, new material being added to it at the spreading ridge by differentiation of the pyrolite which is a hypothetical mixture of 3—5 parts of peridotite and 1 part of Hawaiian basalt. The lithosphere slides on the seismic low velocity layer which is weak on account of partial melting. Creep by dislocation can also take place in the upper mantle between depths of 50 and 400 km, that being the region where the temperature comes close to the melting temperatures of the silicates. Below that depth pressure raises the melting temperature at a faster rate, thus preventing flow, although new estimates of the viscosity of the lower mantle are lower than the previously accepted ones, making mantle-wide convection not altogether impossible (Carter and Ave'Lallemant, 1970). A discussion of the mechanical aspects of sea-floor spreading based on laboratory experiments on flow by dislocation in olivine is given by Ave'Lallemant and Carter (1970). In oceanic areas the depth to the top of the low-velocity layer is about 50 km, while under the continents it is about 150—200 km deep. The top of the low-velocity layer can be regarded as an isothermal surface which is shallower under the oceans than under the continents, so that a plate consisting of both oceanic and continental parts can move with ease on the low-velocity layer (asthenosphere) without disturbing the isotherms (Sclater and Francheteau, 1970). The cold lithosphere descending under the island arc at an angle of 45° or more produces andesitic volcanism on the island arc and sometimes high heat flow on the floor of marginal seas behind it from the rise of liquid magma like in the case of Seas of Japan and Okhotsk (Hasebe et al., 1970), as indicated by Ringwood (1969) in Fig.65. That the descending plate is cold is attested by anomalously low attenuation of elastic waves from earthquakes (Isacks et al., 1968). But the most convincing evidence for the descent of the cold plate comes from plotting the position of earthquake foci projected on a vertical plane perpendicular to the ocean trench. The foci lie on a surface dipping under the island arc, often called the Benioff zone after its discoverer.

FOCI OF EARTHQUAKES

Foci location is now so precise that the actual shape of the descending plate and even perhaps its fall to the bottom of the asthenosphere are outlined by the position of the foci, most beautifully illustrated in Fig.66. The ability of the descending plate to generate

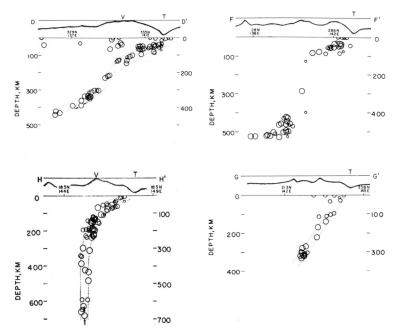

Fig.66. Earthquake hypocenters outlining subducting plates. *DD', FF', HH'* are vertical sections perpendicular to the Izu-Bonin Trench. *GG'* is a section perpendicular to the northern part of the Mariana Arc. *T* denotes trench; *V*, volcanic chain. Circles indicate earthquake hypocenters within 100 km of section. Larger circles denote more precise hypocentral locations. Converging arrows indicate axis of maximum compression, diverging ones axis of maximum dilation for earthquake mechanism solutions. Bathymetry shown with a vertical exaggeration of about 10. (After Katsumata and Sykes, 1969.)

deep earthquakes (Isacks and Molnar, 1971) depends on its brittleness which decreases with increasing temperature so that we can expect the average depth of earthquakes to increase with the downward speed of the plate. The speed, however, cannot be determined seismically; one has to estimate it from the rate of sea-floor spreading as determined by the magnetic anomaly pattern of the associated ridge or ridges. Fig.67

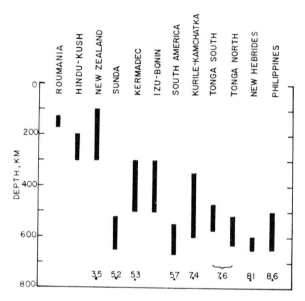

Fig.67. Depth range of maxima in the seismic activity (numbers of earthquakes) as a function of depth in island arcs and arc-like structures for which data are sufficiently numerous. The numbers at the bottom of the figure give the rate (in centimeters per year) of convergence for the arc. Note that the maxima occur over a wide range of depths and that the depths appear to correlate, in general, with the calculated slip rate. (After Isacks et al., 1968.)

shows the most remarkable dependence of depth of earthquake foci on the rate of subduction determined in this way usually from data thousands of km away. We can generalize by saying that just as lineated magnetic anomalies are associated with sea-floor spreading, so deep-focus earthquakes and volcanism indicate consumption of the lithosphere whether it is occurring or has occurred within the last 10 m.y. (the estimated time constant for earthquake production of the plate), on land or in the ocean. The mechanism provides a simple explanation for deep-focus earthquakes which have been tantalizing geophysicists for many decades: the deep earthquakes occur from brittle failure at about the same temperature as the shallow ones.

OTHER EVIDENCE FOR CONSUMPTION

Magnetic evidence for consumption of the lithosphere at oceanic trenches consists of the disappearance of the lineated magnetic anomalies where the latter reach the trench. This is dramatically illustrated south of the Aleutian Trench by the closely spaced magnetic profiles of Fig.68 and 71. Downward dipping ocean sediments toward the trench as revealed, for example, by seismic profiling off Barbados in the Atlantic (Fig.69) and off the Chile Trench (Fig.70) constitutes another independent line of evidence that supports underthrusting of oceanic lithosphere at deep trenches whether or not the motion is toward a continent like South America, or just an island arc like the Antilles with another oceanic plate behind it. On the continents, the Himalayas can be pictured as caused by the underthrusting of the Indian plate under the Eurasian plate, and the Andes by the underthrusting of the southeast Pacific plate under the South American plate.

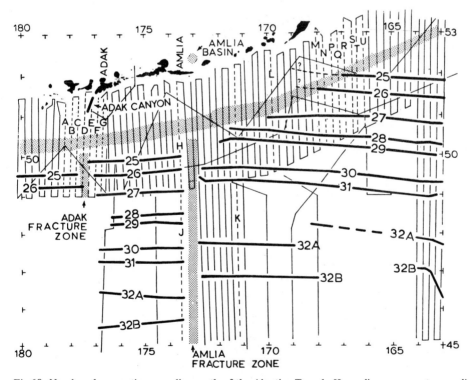

Fig.68. Numbered magnetic anomalies south of the Aleutian Trench. Heavy lines represent anomalies; lighter lines idealized track lines. The age of anomalies increases to the south. Shaded areas denote fracture zones, Aleutian Trench below 6,000 m depth, and the Amlia Basin. (After Grim and Erickson, 1969.)

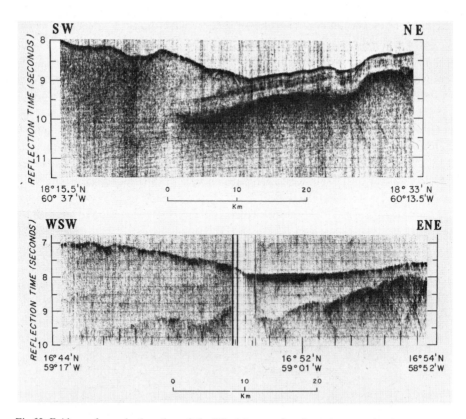

Fig.69. Evidence for underthrusting of the lithosphere under the eastern margin of the Antilles Island Arc in a continuous seismic profiling record. (From Chase and Bunce, 1969.)

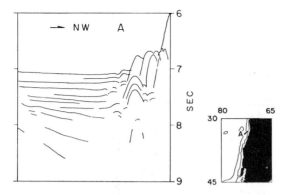

Fig.70. Seismic profiler section and location chart in the coastal area of Chile suggesting under-thrusting and pilung up of marine sediment from subduction of the ocean floor under the South American plate. What appears as pile-up of marine sediment can also be interpreted as sediment slumps from the continent. (Ewing et al., 1969.)

TRANSFORM FAULTS AND EPICENTERS

   Bathymetric and magnetic surveys have shown that expanding oceanic ridges consist of
straight sections nearly parallel to each other which are connected by faults at right angles
to them. Each section of the ridge terminates abruptly at these "transform faults"
(Wilson, 1965). As the new ocean floor with its tell-tale magnetic anomalies is extruded
from the ridge segments, sliding takes place along the transform faults at twice the
spreading rate as indicated on the diagram of Fig.72A. Had the ridge been at one time
continuous and then split into two sections by motion along a transcurrent fault illustrated on Fig.72B, the displacement would be in the opposite direction. The expanding

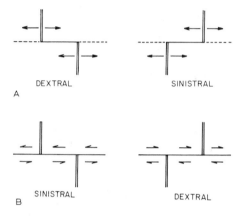

Fig.72. Sense of displacements associated with transform faults (A) and with transcurrent faults (B).
Double line represents crest of mid-ocean ridge; single line, fracture zone. Terms "dextral" and
"sinistral" denote sense of motion on active portions of faults. (From Sykes, 1968.)

ridges and the slipping transform faults are clearly outlined by shallow earthquake
epicenters. Furthermore, for some earthquakes, the orientation of the fault plane and the
direction of motion in that plane producing the earthquake can be calculated from the
magnitude and direction of the first motion registered at several stations well distributed
in azimuth around the epicenter. In every case the direction of slip computed from
earthquake-mechanism solutions agreed with the direction expected for a transform fault
illustrated in Fig.72A. These earthquake-mechanism solutions can also distinguish vertical
from horizontal faulting. The precision with which shallow earthquakes can delineate
ridges and transform faults is illustrated in Fig.73–75.

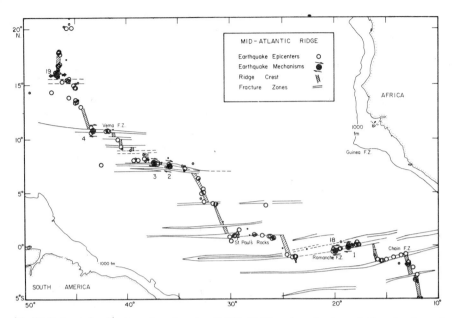

Fig.73. Relocated epicenters of earthquakes (1955–1965) and mechanism solutions for six earthquakes along the equatorial portion of the Mid-Atlantic Ridge. Ridge crests and fracture zones from Heezen et al. (1964a,b, quoted in Sykes, 1968). Events, *1, 2, 3, 4,* and *18* were characterized by a predominance of strike-slip faulting; sense of shear displacement and strike of inferred fault plane are indicated by the orientation of the set of arrows beside each of these mechanisms. (From Sykes, 1968.)

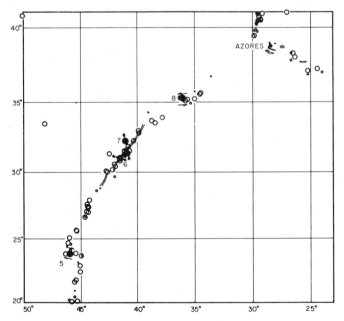

Fig.74. Relocated epicenters of earthquakes along a portion of the Mid-Atlantic Ridge. Rift valley in Mid-Atlantic Ridge is denoted by diagonal hatching. Other symbols same as Fig.73. For events *6* and *7* thick arrows denote the inferred axes of maximum tension. Axis of tension poorly determined for event *7*. (From Sykes, 1968.)

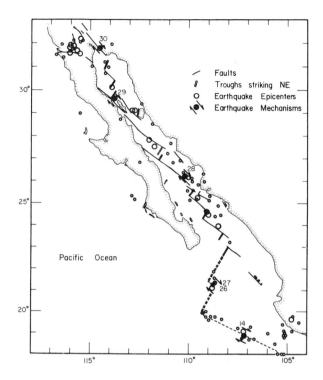

Fig.75. Structural features of the Gulf of California. Earthquake epicenters recomputed for period of 1954–1962. Interpretations of major faults and troughs based on chart of submarine topography of the Gulf of California (Fisher et al., 1964, quoted in Sykes, 1968). Deep troughs striking northeast are interpreted as tensional features that mark sites of growth of new sea floor. (From Sykes, 1968.)

EARTHQUAKES OUTLINE TECTONIC PLATES

Since shallow earthquakes are even more numerous at deep sea trenches and other regions where the lithosphere is consumed, we can say that shallow earthquakes establish present boundaries between the moving plates into which the earth's surface can be divided (Fig.76).

The world map of earthquake-mechanism solutions, presented on Fig.77 gives the direction but not the magnitude of the relative motion between adjoining plates independently of sea-floor spreading which is determined from magnetic anomaly patterns and transform faults. Incipient splitting of these plates is being looked for in the alignment of the few earthquakes that do not fall on the known boundaries like the ones in the Indian Ocean between the tip of India and northwest Australia (Sykes, 1970). On land, the African plate shows signs of splitting in East Africa into a western part called

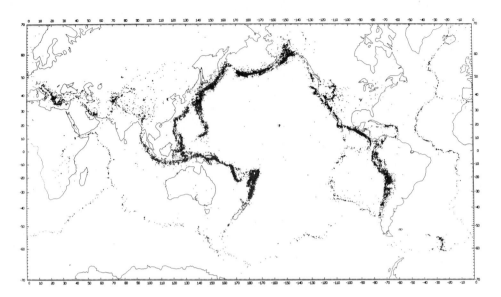

Fig.76. Worldwide distribution of all earthquake epicenters for the period 1961–1967 as reported by U.S. Coast and Geodetic Survey, after Barazangi and Dorman (1968, quoted in Isacks et al., 1968). Note continuous narrow major seismic belts that outline aseismic blocks; very narrow, sometimes steplike pattern of belts of only moderate activity along zones of spreading; broader very active belts along zones of convergence; diffuse pattern of moderate activity in certain continental zones. (From Isacks et al., 1968.)

the Nubian plate and an eastern one called the Somalian plate. It is presumed that because of the greater thickness of the lithosphere and the complexity of the structure of the continents, earthquakes on the continents are more scattered at the plate boundaries than they are in oceanic areas. This scatter is most evident at the borders of the Eurasian plate with the Indian and the African plates (Fig.76).

PLATE ROTATIONS

The relative motions of the plates at the present instant of geologic time can be represented by angular velocity vectors through their poles of rotation (McKenzie and Parker, 1967). Two or more known vectors can be added around a circuit to give the direction and magnitude of a vector that has not been directly measured. This analysis is valid only for infinitesimal rotations. Because lineated magnetic anomaly patterns permit us to follow sea-floor spreading and subduction for at least 80 m.y., the temptation is great indeed to extrapolate some present-day rotations into the geologic past. This lack of rigor is often justified by the great length of some of the transform faults pictured on Fig.82, a fact that implies that some centers of relative rotation of two plates have

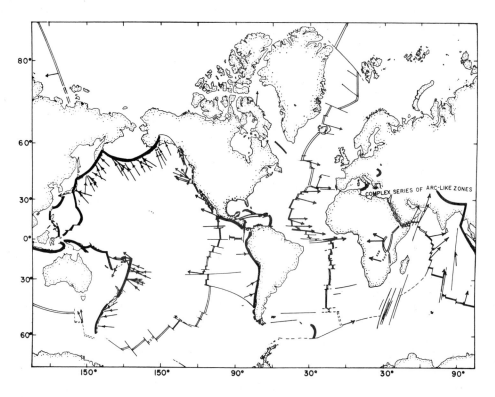

Fig.77. Summary map of slip vectors derived from earthquake-mechanism studies. Arrows indicate horizontal component of direction of relative motion of block on which arrow is drawn to adjoining block. Crests of world-rift system are denoted by double lines; island arcs, and arc-like features, by bold single lines; major transform faults, by thin single lines. Both slip vectors are shown for an earthquake near the western end of the Azores–Gibraltar Ridge since a rational choice between the two would not be made. (Isacks et al., 1968.)

remained relatively fixed for periods of at least as long as the known geomagnetic chronology, making extrapolations into the past useful and excusable.

Consider a block of the earth's surface to crack apart as indicated by shading on Fig.78, block *2* separating from block *1* by rotation about a pole *A*. Any constant relative motion of two such blocks can be represented by rotation about a fixed center. The double lines of the boundary between the plates represent spreading ridges, the single lines transform faults. The ridges recede from the original surface of block *1* by the speed of spreading while the blocks are separating at twice that rate. The transform faults are small circles about pole *A* and should plot as straight horizontal lines on a Mercator projection about that pole. The spreading rate should be proportional to the cosine of the latitude from pole *A*, and can be measured from the magnetic anomaly pattern. If the ridges are active, they can be traced on the chart along with their transform faults from the bathymetry and earthquake epicenters. Great circles perpendicular to the transform faults should intersect at the pole of relative rotation.

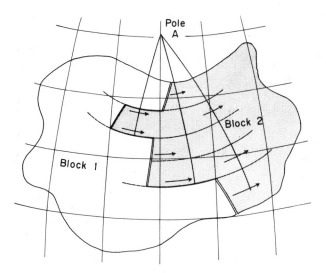

Fig.78. On a sphere, the motion of block *2* relative to block *1* must be a rotation about some pole. All faults on the boundary between *1* and *2* must be small circles about the pole *A*. (From Morgan, 1968.)

EXAMPLES OF PLATE ROTATIONS

An example of this procedure for the Pacific–Antarctic Ridge is shown in Fig.79. The measured spreading rates from a number of magnetic profiles have the expected dependence on distance from the center of rotation (Fig.80).The center of rotation of the Pacific plate with respect to the North American plate can be found by erecting great circles perpendicular to strike-slip faults or from earthquake mechanism solutions in western North America (Fig.17).

The location of poles of relative rotation between two plates can be checked by constructing a Mercator chart about each pole. On such a chart, transform faults should appear as horizontal straight lines. Fig.82 demonstrates this for the presently spreading segments of the East Pacific Rise and in Fig.83 for the fossil fracture zones of the eastern Pacific which are now quiescent because the ridges they were associated with have disappeared. Fig.83 is actually an oversimplification because spreading in the Northeast Pacific has changed direction from time to time, as we shall see later.

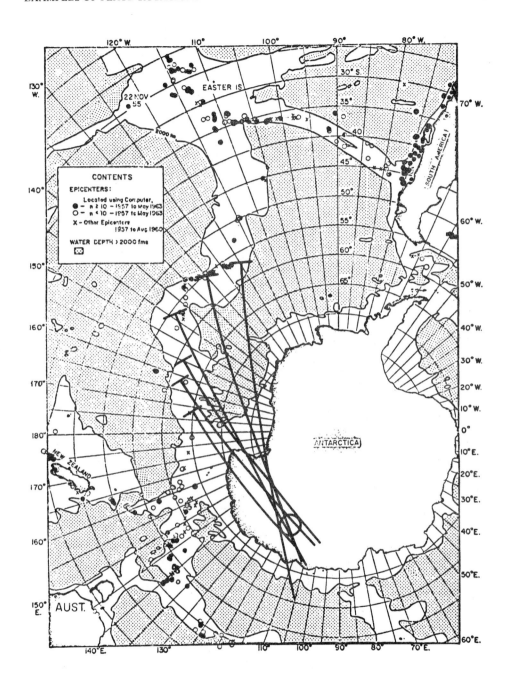

Fig.79. Great circles constructed perpendicular to the strike of fracture zones offsetting the Pacific-Antarctic Ridge are plotted on Sykes's (1963) seismic map of this region. The great circles all pass within 2° of the pole at 71°S 118°E. (From Morgan, 1968.)

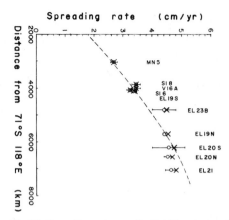

Fig.80. Spreading rates on the Pacific–Antarctic Ridge are compared with a model with V_{max} = 5.7 cm/year about a pole at 72°S 118°E shown as the dashed curve. The circles are the spreading rates measured perpendicular to the strike of the ridge; the crosses are these rates projected parallel to the direction of spreading. (From Morgan, 1968.)

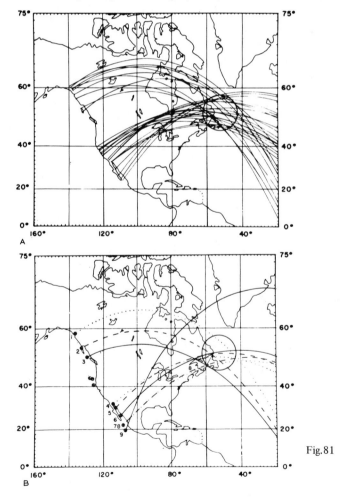

Fig.81

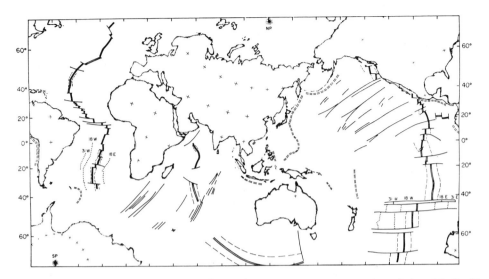

Fig.82. Mercator map projected about 69°N 157°W as the pole. Active transform faults of the Pacific project as straight horizontal lines. Single dashed lines designate numbered magnetic anomalies. Double lines indicate ridge segments; double dashed lines deep trenches. (From LePichon, 1968.)

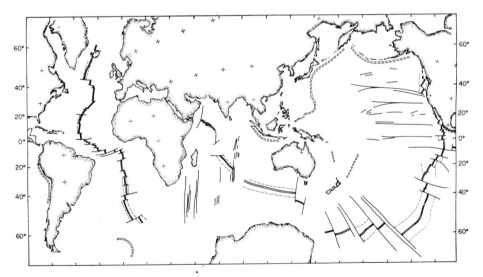

Fig.83. Mercator map projected about 79°N 111°E the center of spreading for the "fossil" Pacific fracture zones. From LePichon (1968). A more detailed analysis by Francheteau et al. (1970) showed considerable mobility of the pole and a different general location. Compare Fig.98 and 99. (From LePichon, 1968.)

Fig.81.A. Composite of great circles perpendicular to selected segments of faults from four separate regions of the west coast of North America. B. Great circles constructed perpendicular to strikes determined from earthquake-mechanism solutions. The circle of intersection drawn has the coordinates 53°N ±6°, 53°W ± 10°. (From Morgan, 1968.)

CHAPTER 7

ABSOLUTE DISPLACEMENT WITH REGARD TO THE EARTH'S ROTATIONAL AXIS

Earthquakes and the magnetic record of recent sea-floor spreading provide only the relative present-day motions of the plates constituting the earth's surface. An area comprising, say, two plates rotating about a common center can be moving as a whole with respect to geographic coordinates. Methods which measure absolute displacements in latitude and orientation with respect to geographic north are used to supplement the information from plate tectonics for reconstructing the geographic position of continents and oceans in the geologic past.

PALEOMAGNETISM OF SEAMOUNTS

The principal method for doing this is paleomagnetism. The direction of magnetization of oriented rock specimens collected from land situated on the same plate as the ocean bottom under investigation can be used for describing a likely motion it could have suffered to come to its present position. Paleomagnetism of seamounts can also be used where land rocks are unavailable or to supplement other information. As this method is not generally known, it will be described in some detail.

By combining magnetic and bathymetric surveys, the magnitude and direction of the magnetization of the seamount can be calculated, assuming that the magnetization is uniform and that the seamount rests on a horizontal plane. The assumptions are tested by recalculating the magnetic anomaly from the computed magnetization, then comparing it with the observed anomaly point by point. The interpretation of the direction of magnetization is carried out by standard paleomagnetic methods, and requires an age determination from a K—Ar age of a rock dredged from the seamount or from some small fossils found in the vescicles of the dredged basalt. If dated lineated magnetic anomalies are present, they give a maximum age to the seamount. Usually the age is not as well known as the magnetization. The original calculation is presented on Fig.84 because it is easier to understand than the one which superceded it (Talwani, 1965) for reasons of greater flexibility and economy of computer time. The seamount is divided into elementary rectangular parallelopipeds the facets of which face magnetic north, magnetic east and the vertical. The uniform magnetization I consists of components A, B, and G. As we saw in the opening chapter, uniform magnetization can be replaced by concentrated poles at the center of the elementary surfaces at distances r_α, r_β, r_γ from the total intensity magnetometer which is titled by the angle i from the horizon. We shall assume

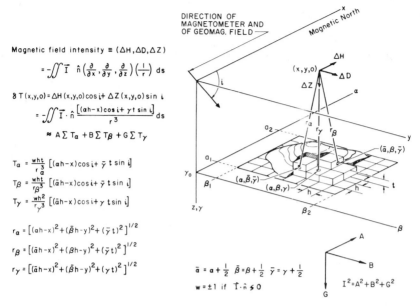

Fig.84. Calculation of the magnetization of a seamount from combining topographic and magnetic surveys. (From Richards et al., 1967.)

here that the anomalous field does not appreciably deflect the magnetometer so that we can neglect the anomaly in declination ΔD. The anomaly is thus approximately equal to:

$$\delta T = \Delta H \cos i + \Delta Z \sin i$$

The contributions T_α, T_β, and T_γ of the three species of poles are calculated from simple geometry. We thus get a number of equations equal to the number of field observations of ΔT from which the components of magnetization A, B and G are computed by least squares. Having obtained these quantities, the anomalous field δT is computed from them for comparison with the observed field ΔT. At each point the difference $|\Delta T - \delta T| = R$ is an individual error which can be mapped. An overall evaluation of accuracy can be obtained from the ratio:

$$r = \frac{\Sigma |\Delta T|}{\Sigma |R|}$$

The disadvantage of this measure is that it gives equal weight to all points.

A seamount survey starts with approximately locating its top by a series of rectangular maneuvers. When satellite navigation is not available, a radar reflector buoy is anchored

Fig.85. Survey of seamount Z-4-2 at 22°22′N 148°14′E. A. Topography (depth) in meters. B. Observed field in gammas. C. Input field. D. Computed anomaly. E. Anomaly field. F. Residual field. C-F in milligauss. Data used for calculation is contained within outer limits of stipled area. (From Vacquier and Uyeda, 1967.)

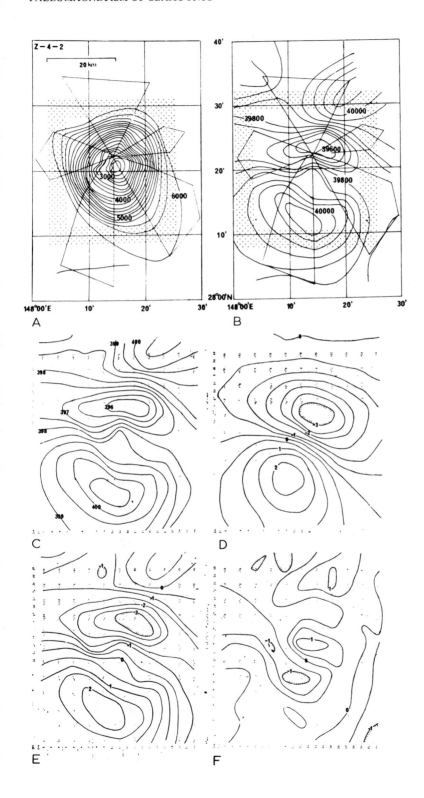

near the summit by a slack line, (usually a 1/4-inch diameter polypropylene rope). To be effective, the reflector should be mounted on a pole not less than 7 m high. A cheap expendable buoy can be made of bamboo pole and a styrofoam plastic float ballasted by anchor-chain links. A small battery-powered light and a flag help to sight the buoy when it is close to the ship. The survey usually consists of a star-shaped pattern (Fig.85) with the buoy in its center. Reflection from the buoy is lost at a range of about 12 nautical miles, so that on the outer part of the pattern the ship is located by dead reckoning corrected for drift.

Fig.85C is the contour map of magnetic intensity values interpolated at the corners of the coordinate grid from the observed magnetic map. Fig.85E is the input magnetic anomaly chart obtained from Fig.85C by subtracting a constant and a uniform magnetic slope which are calculated by the program. Fig.85D gives the anomaly recalculated from the computed components magnetization A, B, and G. Finally, the residual R, which is the difference between Fig.85E and Fig.85D is contoured on Fig.85F. Probably the largest contribution to the residual is lack of uniformity of magnetization, for even if the lava of the shield volcano is uniform in composition, the shape of the bottom surface may depart from a horizontal plane. Furthermore, it is conceivable that the growth of the seamount has straddled one or more inversions of geomagnetic polarity. The latter possibility may explain why old seamounts have given determinations of acceptable quality more often than the younger ones, since geomagnetic reversals were less frequent during the older epochs (Fig.46).

In interpreting the measurement of magnetization we assume that the magnetization induced by the present geomagnetic field is negligible in comparison with the remanent one. This is justified by the high value of the ratio of remanent to induced magnetization from the measurements of rock specimens quoted in Appendix 3.

CALCULATION OF VIRTUAL GEOMAGNETIC POLES

The paleomagnetic measurements on seamounts are treated by the same methods as measurements on rocks collected on land, except that it is impossible to remove the unstable part of the magnetization. Because of the small particle size as revealed by low susceptibility, it is likely that most of the magnetization is thermoremanent. To combine paleomagnetic measurements from different geographic locations one computes the position of the virtual geomagnetic pole for each site, tacitly assuming that the non-dipole field had 2,000–10,000 years to average itself out so that rocks at all epochs have been magnetized by the field of an axial geocentric dipole. Magnetization of oceanic sediments in Fig.39 supports this view. The calculation is given in Fig.86.

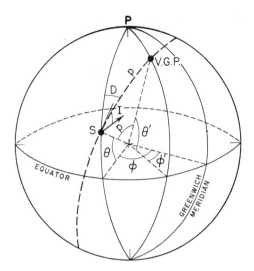

Fig.86. Relationships between location of seamount or sampling site (S), field direction (I, D), and virtual geomagnetic pole (V.G.P.). θ is the latitude and ϕ the longitude of the site; θ' is the latitude and ϕ' the longitude of the virtual geomagnetic pole; D is the declination and I the inclination of the field direction; p is the geomagnetic co-latitude. The coordinates of the V.G.P. are computed from: $\tan I =$ $2 \cot p$; $\sin \theta' = \sin \theta \cos p + \cos \theta \sin p \cos D$; $\sin (\phi' - \phi) = (\sin p \sin D)/\cos \theta'$.

THE POLAR CURVE

In present geographic coordinates, the path by which the virtual geomagnetic pole (V.G.P.) at a certain epoch has taken from some remote epoch is called the "polar curve". It is produced by a gradual change in position of the site with respect to the earth's rotational axis, and one cannot distinguish whether the earth's crust has slipped as a whole with respect to the earth's axis, or whether ít was only due to the motion of the plate on which the site is located, or whether perhaps both processes are responsible for the shift. The contributions of the two processes can be evaluated only if measurements are carried out at many sites widely separated in longitude. The two sources contributing to the polar curve of a plate are called in the literature *polar wander* and *continental drift*.

Irving (1964) defines polar wander: "For the present purposes, therefore, the hypothesis of polar-wandering is thought of as incremental movements of the coincident spin and magnetic axes relative to the earth as a whole, and *relative continental displacement* or *continental drift* as horizontal movements of land masses relative to one another. The former is a general phenomenon and would be reflected in the paleomagnetic results everywhere. The second involves special movements of each continent and may be expected to cause differences in the paleomagnetic results from different continents.

Mechanically, polar-wandering is thought of as movement of the whole earth relative to the axis of rotation, which, aside from precession and nutation, remains fixed relative to the sun. It seems that this could arise from redistribution of mass due to geological processes in the crust or convection in the mantle."

Unfortunately the curve of the V.G.P. is sometimes called *polar wandering curve* creating understandable confusion. To date all departures of paleomagnetic measurements from the orientation of the present geomagnetic dipole field can be interpreted as due to plate motions. There are no data that conclusively demonstrate polar wandering.

NORTHWARD DRIFT OF THE PACIFIC PLATE

Most of the seamount surveys were made in the Pacific Ocean. From west to east on Fig.87 *J* designates two groups of seamounts called "Japanese", *D* is Dixon Seamount, *MA* is Midway Atoll, *H* is a group of seamounts called "Hawaiian", although their age is

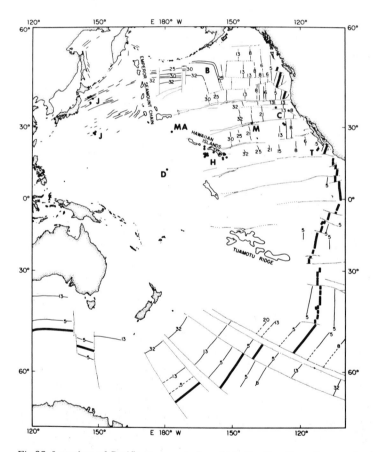

Fig.87. Location of Pacific seamounts for which the direction of magnetization has been computed. Thick line denotes spreading ridge segments, thin lines magnetic anomalies numbered according to the scheme of Heirtzler et al. (1968), Fig.41. Dotted lines denote transform faults. Most of the seamounts are grouped and the nomenclature for them is: *J* = Japanese; *D* = Dixon; *MA* = Midway Atoll; *H* = Hawaiian; *M* = Moonless; *C* = Californian; *T* = Tripod. *B* stands for the Great Magnetic Bight. It is not a seamount. (From Francheteau et al., 1970.)

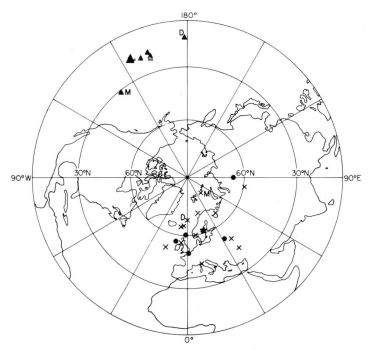

Fig.88. Pole positions for Hawaiian seamounts. Seamount locations marked by triangles. Dixon sea-mount shown by a *D*. Crosses represent north poles, and solid circles represent south poles. The star is the mean pole position excluding the pole for Moonless seamount *M*. Lambert equal-area projection. (From Francheteau et al., 1970.)

Late Cretaceous, *B* is the Great Magnetic Bight, *M* is a seamount belonging to the Moon-less Mountains, *C* is a group called "California", and *T* a group called "Tripod" after an expedition of the same name. It is likely that the Dixon seamount belongs to the "Hawaiian" group. Unfortunately, measurements are lacking in the southern part of the Pacific plate. The V.G.P. for the Hawaiian group shown in Fig.88 scatter just about as much as poles from paleomagnetic measurements on land rocks. The positions of the V.G.P. for the different groups of seamounts and Midway Atoll which appear with their circles of 95% confidence in Fig.89 clearly depend on their age. The "Japanese" group is of Cretaceous age (85 m.y.) from both paleontological and radiometric evidence. So is the "Hawaiian" group (Dymond and Windom, 1968). The "California" seamounts are located between anomalies No.11 and 15 which gives them an Early Oligocene maximum age (40 m.y.). The position of the pole from the Great Magnetic Bight lies about 20° from the present geographic pole obtained by Vine and Hess (1970) by a method described in Fig.93, along the dashed latitude circle as indicated by *B* in Fig.89. The "Tripod" sea-mount group lies between anomalies No.5 and 6 and so are too young to show a signifi-cant displacement. If we suppose that the V.G.P. of the "Hawaiian" seamounts moved up to the North Pole via Midway Atoll by rotation of the North Pacific plate about a fixed center, the position of this center is found approximately by the procedure illustrated on

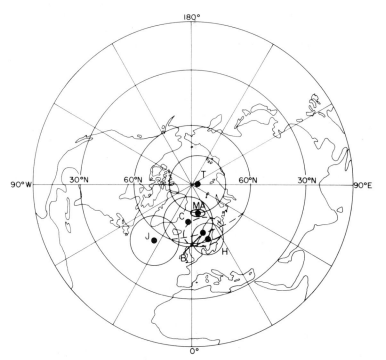

Fig.89. Summary of the mean paleomagnetic pole positions for the Pacific Ocean. T = Tripod sea-mounts; MA = Midway Atoll; C = California seamounts; H = Hawaiian seamounts; locus of the Great Magnetic Bight B is a dashed line (the black circle on the dashed line corresponds to a declination of $3.5°$); J = Japanese seamounts. The ovals surrounding the poles define the 95% confidence limits in the position of these poles. The oval for the bight is a dashed line. Lambert equal-area projection. (From Francheteau et al., 1970.)

Fig.90. A great circle is drawn through points equidistant from the V.G.P. and the North Pole, indicated as a dashed line in the figure. The center of rotation can be located anywhere on this circle. Another such circle is drawn which is equidistant from the North Pole and the V.G.P. of Midway Atoll. The intersection of the two circles gives the center about which the North Pacific plate has rotated toward the North Pole since Cretaceous time. Because the V.G.P. of the "Japanese" seamount group is different from the V.G.P. of the "Hawaiian" group at the 95% level of significance it is likely that the two groups have moved apart by approximately $25°$ in longitude in addition to having traveled about $30°$ northward. That means that the Pacific plate consisted of two parts which have moved apart and then coalesced. Unfortunately this separation in longitude must have occurred in equatorial latitudes, because it left no magnetic record on the ocean floor[1]. This rather poor example illustrates an important feature of paleomagnetic interpretation

---

[1] Note added in proof: Surveys of ten additional seamounts moved the "Hawaiian V.G.P." closer to the "Japanese V.G.P." so that their circles of 95% confidence overlap, making it less likely that the Pacific plate was in two pieces since the Cretaceous (V. Vacquier and C.G.A. Harrison, Paleomagne-tism of Seamounts. *Trans. Am. Geophys. Union.*, April 1972, p.351).

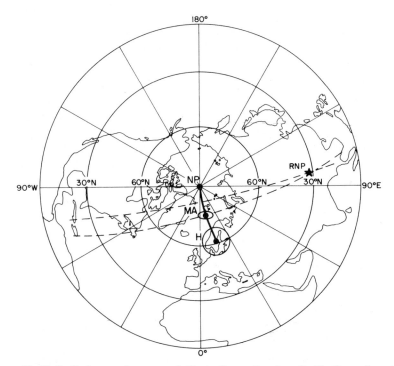

Fig.90. Preliminary polar curve relative to the northeastern Pacific Ocean from the Cretaceous to the present (heavy line). The symbols are the same as for Fig.88. The light dashed lines are great circles defining the pivot point *RNP* represented by a star (see text for explanation) at 33°N 97°E. Lambert equal-area projection. (From Francheteau et al., 1970.)

which might be overlooked, namely that polar curves constructed for two plates can define the relative drift in longitude of one plate with respect to the other. Irving (1964) and Francheteau (1970) discuss examples of the application of this method to paleomagnetic data from continents.

Francheteau et al. (1970) lists 56 paleomagnetic seamount calculations as of December 1969. An outstanding feature of this compilation is that normally magnetized seamounts are more numerous than reversely magnetized ones. If viscous magnetization is to blame for this disparity, the average magnitude of the reversed magnetization should be smaller than the average of the normal one. However, the number of the data is insufficient for a meaningful answer from statistics because of the large inherent variability of the magnitude of the magnetization of both signs. If a portion of the magnetization of North Pacific seamounts is viscous, the distance they have traveled since they were formed is increased.

LATITUDE FROM MAGNETIZATION OF SEDIMENTS

A change in latitude can also be detected by the change in the angle that the remanent magnetization of an ocean sediment makes with the vertical. This method should be

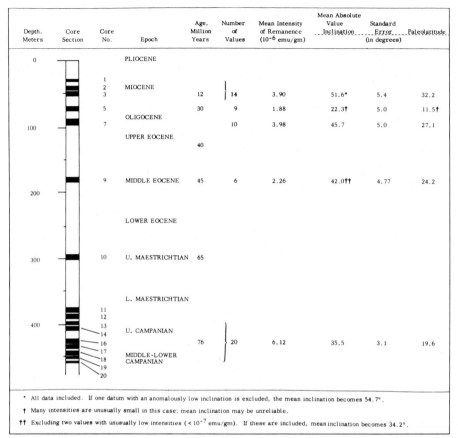

Depth, Meters	Core Section	Core No.	Epoch	Age, Million Years	Number of Values	Mean Intensity of Remanence (10^{-6} emu/gm)	Mean Absolute Value Inclination	Standard Error (in degrees)	Paleolatitude
0			PLIOCENE						
		1							
		2	MIOCENE						
		3		12	14	3.90	51.6*	5.4	32.2
		5		30	9	1.88	22.3†	5.0	11.5†
			OLIGOCENE						
100		7			10	3.98	45.7	5.0	27.1
			UPPER EOCENE						
				40					
		9	MIDDLE EOCENE	45	6	2.26	42.0††	4.77	24.2
200									
			LOWER EOCENE						
300		10	U. MAESTRICHTIAN	65					
			L. MAESTRICHTIAN						
		11							
		12							
400		13	U. CAMPANIAN						
		14							
		16		76	20	6.12	35.5	3.1	19.6
		17	MIDDLE-LOWER						
		18	CAMPANIAN						
		19							
		20							

   * All data included. If one datum with an anomalously low inclination is excluded, the mean inclination becomes 54.7°.

   † Many intensities are unusually small in this case; mean inclination may be unreliable.

   †† Excluding two values with unusually low intensities ($< 10^{-7}$ emu/gm). If these are included, mean inclination becomes 34.2°.

Fig.91. Inclination of natural remanent magnetization as a function of age and depth of deep sea cores from hole No.10, Leg 2 of JOIDES drilling program located at 32°52'N 52°13'W. The change in magnetic inclination corresponds to a curve from 20° – 32°N latitude which is consistent with the polar curve for North America (Irving, 1964), suggesting a rigid connection between the northwestern Atlantic and North America for the last 80 m.y. (From Sclater and Cox, 1970.)

applicable at latitudes below 35° where the dip is sensitive to latitude. Fig.91 presents measurements made on JOIDES Core 10, Leg 2, taken at 32°52'N 52°13'W in the Atlantic. By interpreting the shallowing of the dip angle with depth as caused by a northerly drift of the ocean floor of 2 cm/year since Cretaceous time and comparing this result with paleomagnetic measurements on land, it appears that the northwestern Atlantic was rigidly connected to North America for the last 80 m.y.

     🌲

PALEOLATITUDE FROM ANOMALY SHAPE AND AMPLITUDE

The shape of the magnetic anomalies in the standard sequence is sensitive to latitude as we saw in Fig.51. This enabled McKenzie and Sclater (1971) to postulate a northerly drift

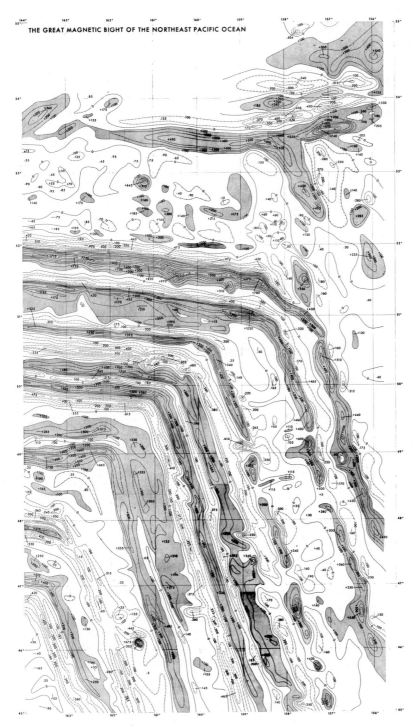

Fig.92. The Great Magnetic Bight of the Northeast Pacific. Residual total magnetic intensity. Shaded areas: > 100 γ. (From Elvers et al., 1967.)

of a part of the Indian Ocean floor (Fig.107). In the Northeast Pacific, magnetic anomalies from No.25 to No.32 of the standard sequence abruptly change direction from 161° to 279°, in what is called the "Great Magnetic Bight" (Fig.92). Vine and Hess (1970) have calculated model anomalies at the present latitude (50°N) and found a smaller ratio of amplitudes of the two sides than the observed one (Fig.93). To agree with the observed amplitude ratio the model must be magnetized at a latitude 20° south of the present location of the bight.

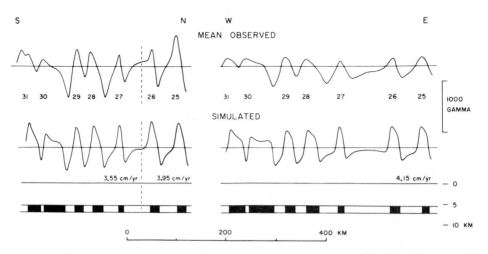

Fig.93. Mean total magnetic intensity anomaly profiles observed on the north–south and east–west branches of the Great Magnetic Bight of Fig.92 are compared to simulated profiles from the standard geochronological model computed at the present latitude for anomalies Nos.25–31. The difference of the ratio of the amplitudes of the north–south and east–west branches between the observed and the computed profiles is interpreted as caused by northward displacement of about 20° in latitude since formation of the anomalies. (From Vine and Hess, 1970.)

THE BIOGENIC EQUATOR

Lastly there is a method for finding drift in latitude restricted to a narrow band about the equator. Because the cold water upwelling at the equator can support a dense animal population, the rate of sedimentation in a band only about 3° wide is several times faster than in the adjoining ocean. The past position of the equator can be found by the relative thickness of paleontologically dated sediments in holes drilled northward from the present equator, as shown in Fig.94. An average northerly drift of 2 cm/year since Early Miocene time appears to be recorded by the northward thickening of the older sediments, agreeing with paleomagnetic measurements.

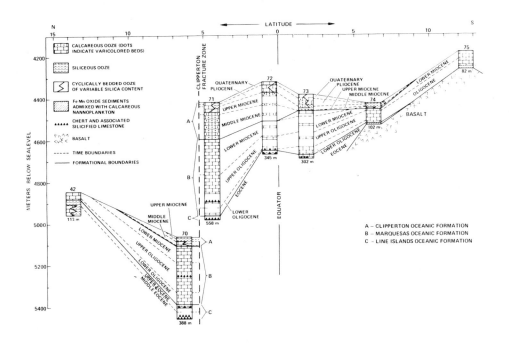

Fig.94. Dated sediment thickness in deep-sea drilling holes situated on a line crossing the equator in the Pacific Ocean. The older sediments first thicken north of the equator then thin out again at 14°N, suggesting a northward drift of the Pacific Ocean floor of 2 cm/year since Early Miocene time. (Tracey and Sutton, 1971.)

CHAPTER 8

INTERACTION BETWEEN THE FLOOR OF THE PACIFIC OCEAN AND ITS MARGINS

The northward drift of the Pacific Ocean floor since Cretaceous time, discussed in the last chapter, has not been caused by polar wandering because the Cretaceous and Early Tertiary V.G.P.'s from paleomagnetic measurements in Europe and America have not participated in this drift (Irving, 1964); in fact the Cretaceous pole from North American rocks lies just north of the Bering Strait at 68.5°N 185°E (Creer, 1970; Francheteau, 1970), 21° on the other side of the North Pole from the V.G.P. of the Hawaiian sea-mounts, making the relative displacement across the North Pole between the North Pacific and American plates at least 50° since about 90 m.y.B.P., which amounts to an average speed of about 5 cm/year.

SUBDUCTION IN ALEUTIAN TRENCH

As we have seen in the last chapter, the Great Magnetic Bight gives independent evidence of the northerly drift from the amplitude ratio of the same anomalies in the north—south and east—west limbs near the bend. The morphology of the bight provides much stronger evidence. The western limb of the bight consists of anomalies which are younger toward the Aleutian Trench which must have engulfed the spreading ridge that created them along with some part of the southern set of the anomaly sequence (Fig.95). The spacing of the east—west anomalies gives a spreading rate of 3.5 cm/year requiring the rate of subduction at the trench to be faster than 7 cm/year in order to absorb the ridge. The uniform rate of 5 cm/year we estimated from the position of Cretaceous V.G.P.'s for the Pacific and North American plates is too slow, making it necessary to postulate a faster rate during the earlier part of this drift.

The geology of the Aleutians contains a suggestion of when the east—west ridge was absorbed by the trench. Grow and Atwater (1970) in proposing this connection, quote a number of papers according to which extensive volcanism and plutonism began in the central Aleutians in Early Miocene about 24 m.y.B.P. From this figure we can calculate the rate of subduction of the Pacific plate into the Aleutian Trench and compare it to the slip along the San Andreas Fault. From the spacing of the visible east—west anomalies of the Great Magnetic Bight, the spreading rate of the ridge was 3.5 cm/year. As the last visible anomaly, No.27 is 67 m.y. old, 1,500 km of ocean floor south of the ridge were consumed in the trench at 6.2 cm/year. This very nearly checks the present San Andreas rate, for although the distance to the rotation center at 53°N 53°W is greater, raising the

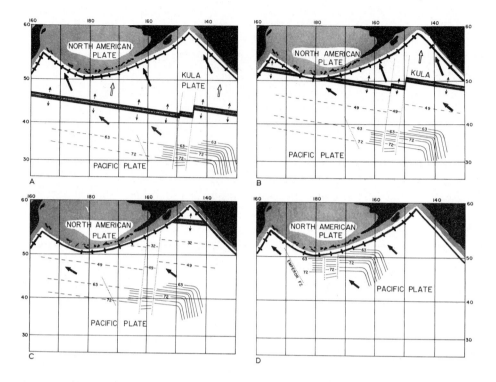

Fig.95. Paleographic reconstruction of the anomalous magnetic field south of the Aleutian Island Arc. A. 34 m.y. ago. B. 22 m.y. ago. C. 16 m.y. ago. D. Present. (Slightly modified from Grow and Atwater, 1970.)

rate by a ratio of 1.2 with respect to the west coast of North America, the motion of the Pacific plate makes an angle of about 54° to the axis of the trench so that only 0.81 of · the velocity is effective. We may conclude that the rotation between the Pacific and North American plates at the central Aleutians has been 7.2 cm/year around the Morgan (1968) pole at 53°N 53°W for the last 24 m.y. or so. This motion is parallel to the western third of the Aleutian Arc where the boundary between the Pacific and the North American plates becomes a transform fault, a deduction supported by lack of active volcanism in the western Aleutians. Volcanism there probably became extinct from the time the east—west ridge disappeared into the trench. However, the persistence of the trench to the present day, long after subduction had ceased, suggests expansion of the western Bering Sea by some other mechanism.

ORIGIN OF THE GREAT MAGNETIC BIGHT

From remaining evidence on the ocean floor and on the continent, it is postulated the bight originated from a junction of three spreading ridges presented in Fig.96. In the

L_____J 500 KM

Fig.96. Postulated configuration of the triple ridge junction from which the anomalies of the Great Magnetic Bight of the Northeast Pacific of Fig.92 originated. Only part of the southwestern pattern is now visible. The rest was swallowed up in the Aleutian Trench and a formerly existing trench along North America. (After Vine and Hess, 1970.)

ocean we see only the southwestern corner of the pattern from anomalies No.25–32. Three plates, the Kula (Grow and Atwater, 1970), the Farallon (McKenzie and Morgan, 1969), and the Pacific plates were growing out of the three ridges. The Kula plate has now completely disappeared under the North American plate as was discussed in detail in the preceding paragraph. Small portions of the Farallon plate are left on the southeast side of the Gorda and Juan de Fuca ridges, which are segments of the East Pacific Rise.

SHIFTING OF SPREADING DIRECTION IN THE NORTHEAST PACIFIC

The very small areas of magnetic lineations east of spreading ridges and the absence of the ridges themselves over most of the northeastern Pacific is most striking (Fig.97). This area is so well-surveyed topographically and magnetically that changes in the direction of spreading appear in some detail (Atwater and Menard, 1970; Francheteau et al., 1970). North of the Mendocino Fracture Zone, as one goes westward, the magnetic lineations swing 40°, from 21°E at the Juan de Fuca Ridge, to 19°W at the Great Magnetic Bight. The great fracture zones of the Northeast Pacific were divided into five sections of same age interval in Fig.98 and the center of relative rotation of the Pacific and Farallon plates were calculated for each epoch from segments of the fracture zones and spreading rates by the method of Morgan (1968) described in Chapter 5. The path of the rotation pole plotted in Fig.99 is far from being smooth, and demonstrates that long transform faults

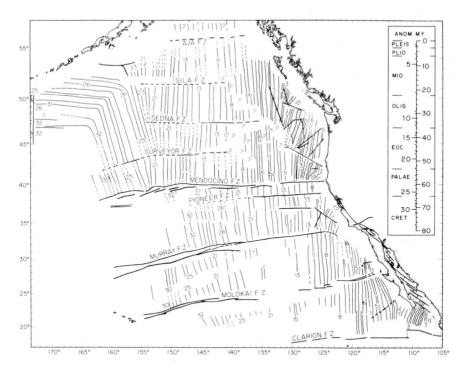

Fig.97. Magnetic lineations and fracture zones in the Northeast Pacific. The Juan de Fuca, Gorda and East Pacific Rise ridges are shown by shading. Anomalies are numbered according to Heirtzler et al. (1968) shown on Fig.41. Note the difference in offsets across the eastern and western portions of the Murray and Molokai Fracture Zones and the anomalously small distance between anomalies 8 and 21 south of the Molokai Fracture Zone. (From Atwater and Menard, 1970.)

fail to stabilize the direction of spreading. Perhaps the frequent shift of the Pacific-Farallon rotation center occurred to compensate for the finite rotations of these plates with the North American plate, which played an important part in shaping the geology of western North America.

Sea-floor spreading at the mouth of the Gulf of California confirms the uniform rate of 6 cm/year between the Pacific and American plates back to about 7 m.y. (Fig.100). Although the central magnetic anomaly is poorly developed, the profiles are fairly symmetrical on the two sides of the ridge. In the following discussion the relative rotation of the Pacific and North American plates will be assumed uniform at 6 cm/year measured at the mouth of the Gulf of California about the fixed center at 53°N 53°W of Morgan (1968).

The hypothetical reconstruction of the plate tectonics in the northeastern Pacific is illustrated in Fig.101 at three stages of development. The hypothetical junction of the three ridges which have generated the Great Magnetic Bight is drawn on the earliest picture as double dashed lines. About anomaly 10 time, 32 m.y. ago, the Pacific plate collided with the North American plate, approximately at the tip of Baja California

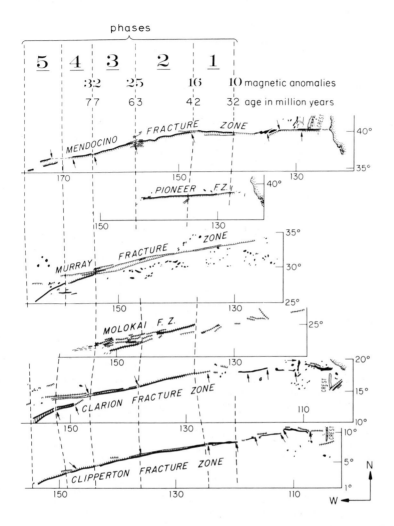

Fig.98. Fracture zones in the northeastern Pacific aligned by the younger of the two anomalies on either side of the fracture zone. The arrows are the changes of direction. The heavy black line represents the sections of the fracture zones used to compute the poles. The Murray Fracture Zone has not been considered for phases 1 and 2. During this time interval it behaved as a leaky transform fault and there was a jump in the spreading centre below the Murray. (From Francheteau et al., 1970.)

(McKenzie and Morgan, 1969). At that time the relative motion of the Farallon plate with respect to the Pacific plate was 10 cm/year from the Heirtzler et al. (1968) chronology. Previously we have seen that the motion between the Pacific and American plates is 6 cm/year at the mouth of the Gulf of California about 53°N 53°W. Adding these velocities (Fig.101A) we get 7 cm/year for the rate of consumption of the Farallon plate in a trench that formed its eastern boundary. This event marks the birth of the transform fault a partial expression of which is the San Andreas Fault. The progress of the widening of the boundary between the Pacific and American plates can be followed on Fig.97 by

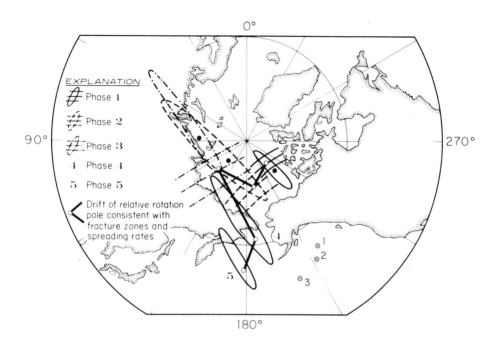

Fig.99. Poles of relative rotation Northeast Pacific/Farallon. Mean and 95% band of confidence for the angular distance of the pole of relative rotation Northeast Pacific/Farallon, are determined from the spreading rates for magnetic anomaly intervals 8– 13, 13– 21, 23– 32. Also shown are the pole (black dot) and oval of confidence for the fracture zone data for magnetic anomaly intervals 10– 16, 16– 25, 25– 32. The heavy line joining white dots represents the drift of the relative pole Northeast Pacific/Farallon consistent with both fracture zone and spreading rate data. The points *1, 2, 3* are the positions of localities used to compute mean angular distance of pole from spreading rate data for phases *1, 2,* and *3*. (From Francheteau et al., 1970.)

the age of the magnetic anomalies that die out against the continent. From the velocity vectors of the plates, the speed of the southern junction relative to the American plate is 1.1 cm/year to the southeast while the northern junction moves 6.0 cm/year to the northwest, making the length of the transform fault grow at 7.1 cm/year. If one assumes a constant speed of spreading from 32 m.y.B.P. to the present day, one gets 2,270 km between the Mendocino Fracture Zone and the mouth of the Gulf of California which checks the measured distance of 2,300 km. The precision of this agreement is probably fortuitous because since anomaly 5 (10 m.y.B.P.) the spreading of Juan de Fuca and Gorda ridges changed direction, and a slowing down of the spreading rate to 6 cm/year from 10 cm/year about 20 m.y. ago is indicated on Fig.43. Furthermore the small pieces of the Farallon plate squeezed between the Pacific and American plates got slightly broken up and rotated as can be seen by close inspection of the original magnetic anomaly chart (Fig.102). Only 350 km of displacement since the Oligocene can be postulated from geological evidence along the San Andreas Fault so that the transform fault under consideration is a wider fault zone consisting of many parallel faults for

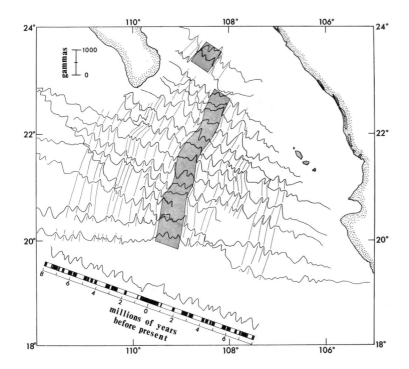

Fig.100. Profiles of magnetic anomalies at the mouth of the Gulf of California. The shaded areas locate the central magnetic anomaly. The magnetic block model and the corresponding anomaly profile were constructed using the geomagnetic time scale of Heirtzler et al. (1968) and a half-rate spreading of 3.0 cm/year. (From Larson et al., 1968.)

taking up the motion together or by turn (Atwater, 1970). Between 10 and 5 m.y. ago (Fig.101D), the San Andreas ridge-transform system jumped from the Pacific Ocean shore into the Gulf of California which initated the present spreading of the gulf. Suddenly, Baja California found itself on the Pacific plate.

Occasional jumping of ridges also has been recorded by magnetic anomalies far from present plate boundaries. If this phenomenon occurs too often in succession, it should become very difficult to recognize the standard magnetic anomaly pattern. It is likely that some oceanic magnetic anomalies have not been deciphered from this cause. A good example of such a jump appears below the Murray Fracture Zone on Fig.97. Anomaly 21 shows a displacement of 680 km across the Murray Fracture Zone, whereas anomaly 13 is displaced only 150 km. Sometime between the epochs of these two anomalies the ridge jumped eastward 530 km.

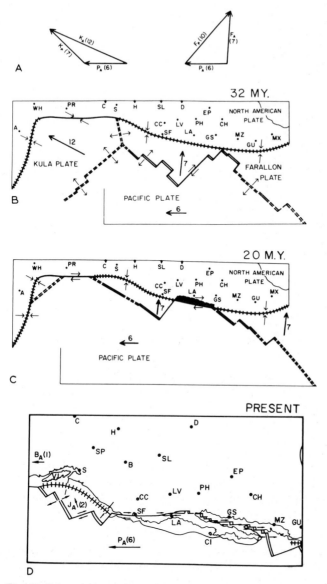

Fig.101. Paleographic reconstruction of plate motions in the Northeast Pacific.

A. Vector diagram of plate motions. $P_A(6)$ stands for Pacific with respect to America at 6 cm/year. Its direction is given by rotation about 53°N 53°W. F stands for Farallon, K for Kula. The direction of Fp is that of the Mendocino Fracture Zone.

B. The hypothetical triple ridge junction which left the Great Magnetic Bight is shown by double dashes. The time of earliest contact of Pacific plate with trench is given by oldest anomaly on coast: 32 m.y.B.P. according to McKenzie and Morgan (1969) and 29 m.y.B.P. according to Atwater (1970). The dots are cities: S = Spokane; SF = San Francisco; LA = Los Angeles; GS = Guaymas; MZ = Mazatlan.

C. The black area is unacceptable overlap which could be diminished by reducing the speed of 7 cm/year or shifting centers of rotation.

D. Note the partial clockwise rotation of the small remnants of the Farallon plate. (Slightly simplified from Atwater, 1970.)

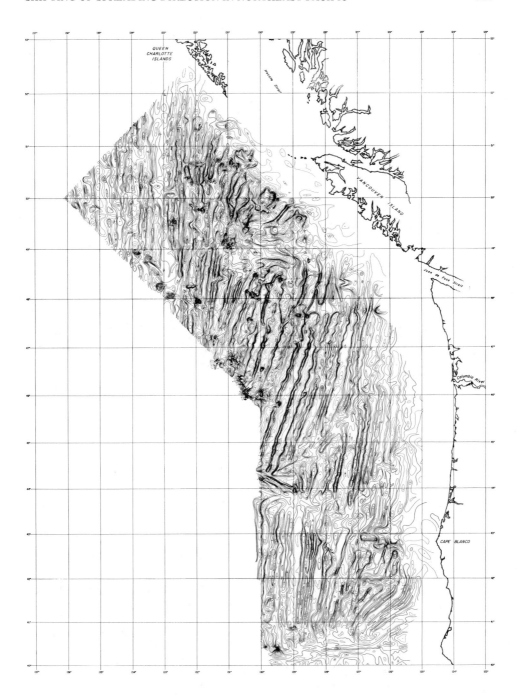

Fig. 102. Magnetic anomaly map off the west coast of North America. Contour interval 100 γ. Magnetic lineations have recorded the breakup, distortion and clockwise rotation of the last remnant of the Farallon plate. (Raff and Mason, 1961.)

CHAPTER 9

THE INDIAN OCEAN

COVERAGE

Although dated magnetic anomalies occupy only a relatively small area of the Indian Ocean, they provided the key information for the reconstruction of its evolution since Late Cretaceous time (Fisher et al., 1971; McKenzie and Sclater, 1971). North of 35°S the Indian Ocean is covered by the relatively dense net of magnetic tracks (Fig.103). The area south of Australia has been left blank on purpose. It is covered in detail by Weissel and Hayes (1971), who have found slightly asymmetrical sea-floor spreading which started 40 m.y. ago and persisted for 20–30 m.y. South of Australia the Southeast Indian Ridge joins the Pacific–Antarctic Ridge. LePichon (1968) calculated the spreading of this

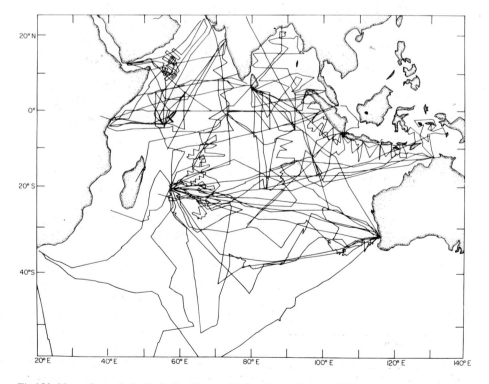

Fig.103. Magnetic tracks in the Indian Ocean. (McKenzie and Sclater, 1971.)

ridge as $6.8 \cdot 10^{-7}$ deg./year around a pole at 36°S 53°E from a finite rotation of −31°. McKenzie and Sclater (1971) got $8.0 \cdot 10^{-7}$ deg./year around a center at 2°S 46°E, a tolerable check as such things go.

TWO SPREADING EPISODES

West of the Ninetyeast Ridge (Fig.104), two periods of sea-floor spreading have been deciphered from the magnetic anomalies whereas east of it and south of Broken Ridge, there appears to have been only one spreading interval which started at anomaly 18 time (46 m.y.B.P.) and which is presently separating Australia from Antarctica. East of Ninety-east Ridge and north of Broken Ridge magnetic anomalies have not been identified. The same holds true for the area south of Madagascar and west of 70°E, despite the numerous earthquakes that define the southwest branch of the Indian Ridge. Three triple junctions show up with unequal clarity on Fig.104. The most obvious one is the junction of the three arms of the Indian Ocean Ridge where the Indian plate meets the Antarctic and the Somali plates. Proceeding north from there along the Carlsberg Ridge we come to its

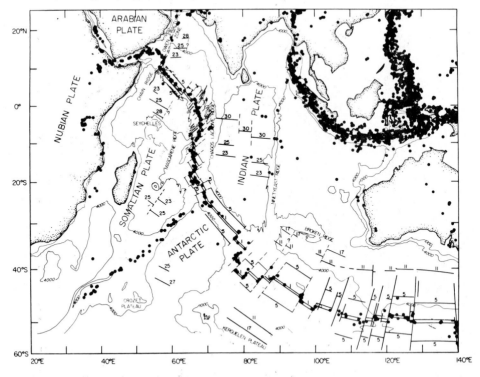

Fig.104. The Indian Ocean Ridge System. Earthquake epicenters associated with ocean floor spreading are shown by dots, ridge crests by double lines and numbered magnetic anomalies by single lines. Depth contours are in meters. (From McKenzie and Sclater, 1971.)

junction with an east–west ridge emerging from the Gulf of Aden and the Owen Fracture Zone which runs parallel to the western shore of the Arabian Sea. At the head of the Gulf of Aden we find a triple ridge junction at the convergence of the Somali, the Arabian and the Nubian plates. The boundary between the Somali and the Nubian plates runs through the East African Rift, which is presumably spreading at a very slow rate of a few mm for the last 10 m.y. (McKenzie et al., 1970). The boundary between these two plates is undefined south of Madagascar. The central part of the Indian Ocean Ridge is broken up into short segments by numerous transform faults. Anomaly 5 (10 m.y. old) runs parallel to the ridge segments. From this anomaly, the ridge crest, and the transform faults, the relative motion between the Indian and the Somali plates comes out $6.2 \cdot 10^{-7}$ deg./year about a center 16°N 48°E. An enlarged picture of the central portion of the ridge presented on Fig.105 shows that the spreading direction is northeast–southwest. The

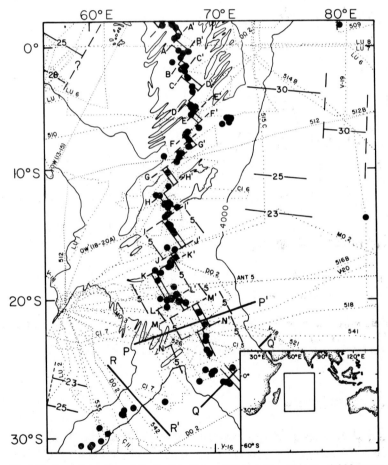

Fig.105. Central portion of the Indian Ocean Ridge outlined by the 4,000-meter contour. Black dots are earthquake epicenters; double lines are segments of the ridge crest; dashed lines are transform faults; single lines are numbered magnetic anomalies; dotted lines are ship and airplane tracks. Bathymetric profiles *PP'*, *QQ'* and *RR'* are shown on Fig.119. (From McKenzie and Sclater, 1971.)

vagueness of the correlations of anomalies other than the central one in the areas immediately adjacent to this latitude interval, is well illustrated on Fig.106. Between 12°S and 2°N anomaly 5 could not be recognized.

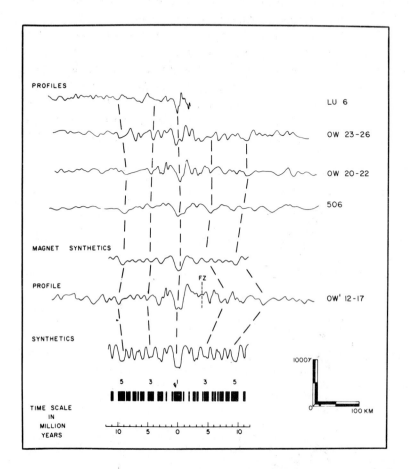

Fig.106. Axial magnetic anomalies across the Carlsberg Ridge between 3 and 6°N. Observed profiles projected onto N45°E. Synthetic profiles generated at right angles to a ridge, striking N45°W, at 5°N with a spreading rate of 1.6 cm/year. (From McKenzie and Sclater, 1971.)

IDENTIFICATION OF OLDER ANOMALIES

The identification of the older anomalies ranging from No.23–30 on Fig.104 was done by comparing the observed anomalies with calculated anomalies from the standard block models formed at different latitudes by the method of Appendix 2. In Fig.107, the shapes of older anomalies in the Arabian Sea seem to fit best the model curve generated at 10°S, suggesting that the floor of the Arabian Sea, east of the Owen Fracture Zone, has

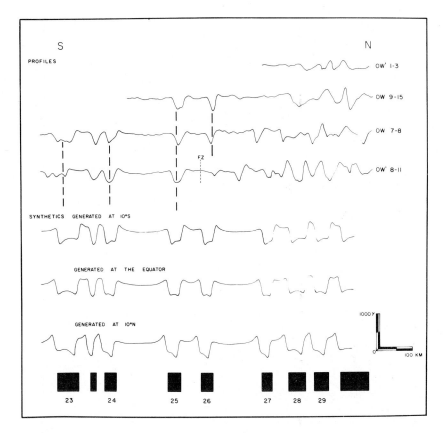

Fig.107. Older magnetic anomalies in the Arabian Sea. Observed profiles projected onto N0°E. Synthetic magnetic profiles generated at right angles to a ridge, striking N75°W, with a spreading rate of 6.5 cm/year observed at 15°N. (From McKenzie and Sclater, 1971.)

moved northward some 25° during the last 70 m.y.

The older anomalies between the Carlsberg Ridge and the Seychelles are compared with models in Fig.108. Here, again, McKenzie and Sclater (1971) believe that the model generated at 10°S fits best the observed anomalies, agreeing with the interpretation of the old magnetic features in the Arabian Sea.

Only the northern set of the old magnetic anomalies was found east of the ridge and south of Ceylon (Fig.104). Comparison with the synthetic profiles in Fig.109 and Fig.110 suggests that these anomalies were generated at 40°S.

The short distances separating anomalies 5 and 23 in Fig.104 are interpreted by Fisher et al. (1970) and McKenzie and Sclater (1971) as an almost complete cessation of spreading between anomalies 6 and 21 and subsequent change in spreading direction which was nearly north—south during the early episode. A part of the angle between the trends of the old anomalies east and west of the ridge comes from rotation about the present center at 16°N 43°E.

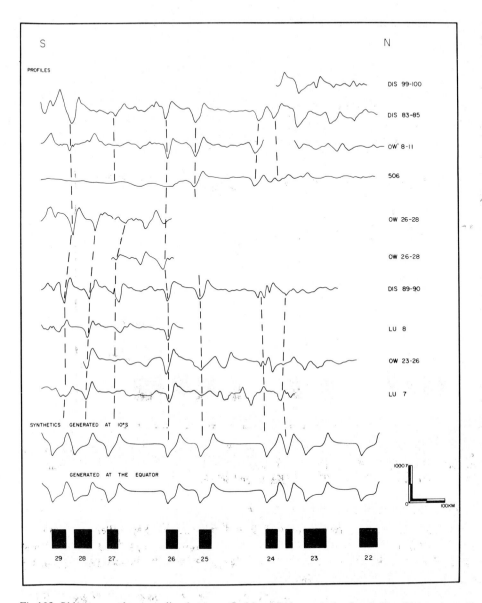

Fig.108. Older magnetic anomalies between Carlsberg Ridge and the Seychelles. Observed profiles projected onto N30°E. Synthetic magnetic profiles generated at right angles to a ridge, striking N75°W, with a spreading rate of 6.5 cm/year observed at the equator. (From McKenzie and Sclater, 1971.)

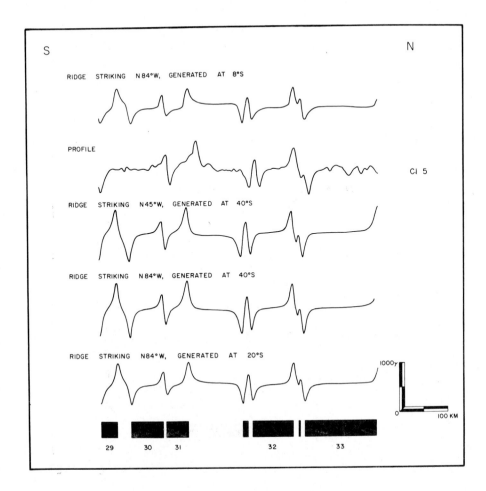

Fig. 109. Comparison between Circe 5 magnetic anomaly profile and two synthetic profiles for the older anomalies. Circe 5 profile projected onto N6°E. The upper synthetic profile generated at right angles to a ridge, striking N84°W at 8°S with a spreading rate of 5.6 cm/year observed at 8°S, with the same spreading rate and observation point. (From McKenzie and Sclater, 1971.)

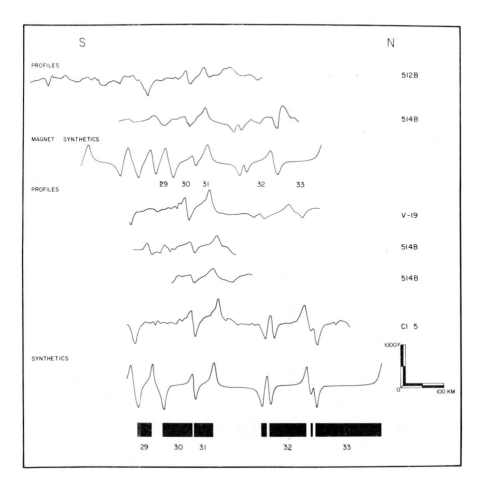

Fig.110. Older recognizable magnetic anomalies in the region south of Ceylon. Observed profiles projected onto N6° E. Synthetic profiles generated at right angles to a ridge, striking N45°W, at 40°S with a spreading rate of 5.6 cm/year observed at 8°S. (From McKenzie and Sclater, 1971.)

EVOLUTION OF THE INDIAN OCEAN FLOOR

The evolution of the ocean floor since the Late Cretaceous is pictured on Fig.111 in three stages. Between anomalies 32 and 21, i.e., between 80 and 50 m.y.B.P., rapid spreading at the half-rate of 5.6 cm/year ($5.0 \cdot 10^{-7}$ deg./year) took place about two east—west ridges separated by a long north—south fracture zone which is none other than the Chagos—Laccadive lineament which was uninterrupted during that epoch. The first diagram, Fig.111A shows the situation at anomaly 21 time 50 m.y.B.P. when this rapid motion ceased or slowed down to an imperceptible rate. It is remarkable that paleomagnetism of Indian rocks gives a pivot point for the rotation of India for 27°N 2°W

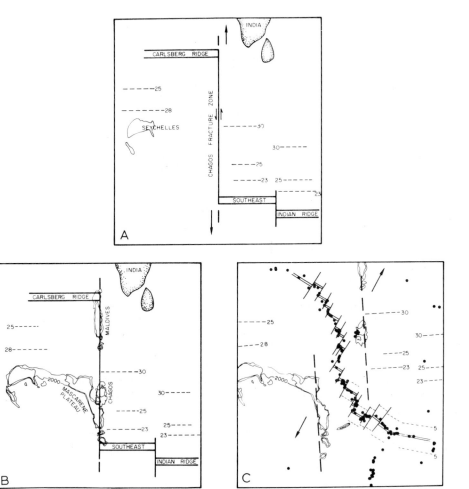

Fig.111. A.Schematic diagram of the central Indian Ocean in Eocene time (anomaly *21*, about 50 m.y.B.P.). B. Schematic diagram of the central Indian Ocean in Miocene time (anomaly *6*, about 20 m.y.B.P.). C. Schematic diagram of structural trends of the central Indian Ridge at the present time. (From Fisher et al., 1971.)

since Jurassic time and an average speed of $10 \cdot 10^{-7}$ deg./year. Furthermore, on a Mercator projection, about the paleomagnetic pivot point of India, the Chagos–Laccadive Fracture is roughly parallel to the edge of the map (Fig.112) as it should be, implying that prior to 50 m.y.B.P. the northward motion of the Indian plate was much faster than that of the Somali plate which also rotated northward about a slightly different center (Creer, 1970). Between 50 and 20 m.y.B.P., the Chagos Bank and the Mascarene Ridge were formed, but the plates did not move appreciably (Fig.111B). At the close of this period the north–south transform fault broke up into the short sections that are forming the present-day spreading by rotation around 16°N 40°E. During the last evolutionary

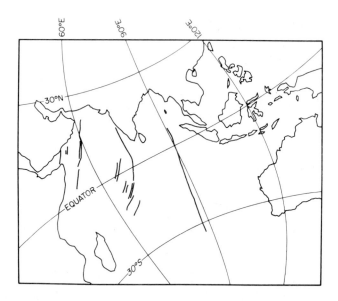

Fig.112. An oblique Mercator projection about the pivot point of India at 27°N 2°W. Only well-substantiated fracture zones in the northwest Indian Ocean are shown. The Ninetyeast Ridge and the Chagos Laccadive Ridge are also shown. Note that the fracture zones north of the Chagos Laccadive Ridge are roughly lines of latitude in this projection. (From Francheteau and Sclater, 1969.)

period, the Chagos Bank and the Mascarene Ridge moved apart to their present positions by rotation about this center, as sketched in Fig.111C where the older anomalies west of the ridge have a substantially different strike from the actual one in Fig.104. This lack of parallelism may imply a more complex history than the one we can presently decipher, and which is hidden in the 30 m.y. hiatus between the two spreading episodes.

CHAPTER 10

DEPENDENCE OF OCEAN DEPTH AND TERRESTRIAL HEAT FLOW ON THE AGE OF THE OCEAN FLOOR

COOLING AFTER EXTRUSION

The lithosphere extruded from a spreading ridge crest gradually cools down as it moves away. Its average temperature drops as a function of its age, which can be determined from the geomagnetic chronology to about 80 m.y.B.P. and from there, by deep-sea drilling of the ocean bottom to the Jurassic, about 200 m.y.B.P.

We may therefore expect terrestrial heat flow to decrease as we move away from active ridges, while the increase of the ocean depth can be interpreted as caused by thermal contraction assisted perhaps by mineral phase changes as the lithosphere is cooling down.

MEASUREMENT OF HEAT FLOW AT SEA

Terrestrial heat flow at sea (Langseth, 1965), is obtained by multiplying the temperature gradient in the first 10 m of bottom mud by its thermal conductivity. A spike or a core barrel carrying temperature sensors is driven into the sediment by a weight containing a recording instrument which measures differential temperatures across distances of about 1 m with a sensitivity of 0.002°C, or better. The thermal conductivity is commonly measured either on a core specimen or in situ (Corry et al., 1968) by the needle probe method of Von Herzen and Maxwell (1959).

AGE BANDS IN THE NORTHERN PACIFIC

The northern Pacific is divided in Fig.113 into nine areas increasing in age from east to west. The distribution of 585 heat-flow values selected for their reliability is shown by open circles. A statistical test demonstrated that within each age band the departures from the mean obeyed the normal distribution law. Heat flow decreases with age as seen on Fig.114, most of the decrease occurring in the first 50 m.y.

CONTINENTAL HEAT FLOW

A similar plot can be made of heat flow vs. age for heat flow on the continents

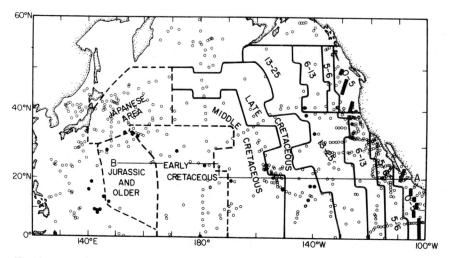

Fig.113. Age provinces of the North Pacific. The boundaries of the provinces are determined from the magnetic lineations (Atwater and Menard, 1970) and the age of sediment recovered by JOIDES deep-sea drillings (McManus and Burns, 1969); Scientiff Staff, Leg VI, (1969). The thick black line is the crest of the East Pacific Rise, Gorda and Juan de Fuca ridges. The open circles are the locations of the heat-flow measurements and the filled-in circles are the locations of the JOIDES deep-sea drillings. (From Sclater and Francheteau, 1970.)

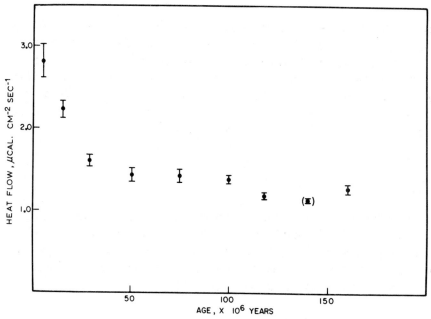

Fig.114. Plot of mean heat flow against age of province for the North Pacific. The length of the bar gives the magnitude of the respective standard error. The mean value for the youngest province has been plotted for a mean age of 5 m.y. to take care of the paucity of observations in the crestal regions. (From Sclater and Francheteau, 1970.)

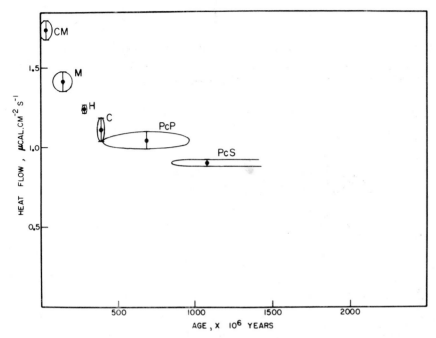

Fig.115. A plot of mean heat flow against age of orogenic province for continents. *CM* = Cenozoic miogeosynclines; *M* = Mesozoic folding; *H* = Hercynian orogeny; *C* = Caledonian orogeny; *PcP* = Precambrian platforms; *PcS* = Precambrian shields. (After Polyak and Smirnov, 1968; Smirnov, 1968, fig.1.)

(Fig.115). It is likely that a correction of +10 to 15% should be applied to the continental heat-flow values because many of the conductivity measurements were made on dry specimens. It takes in the neighborhood of 1,000 m.y. for a rock unit on land to reach the equilibrium value of heat flow. The disparity of the time scales for the decrease of heat flow on land and at sea shows that a large portion of the heat flow in the two regions originates from different causes.

For something like 15 years the approximate equality of terrestrial heat flow at sea and on land of $1.5 \pm 10\%$ μ cal.cm^{-2} sec^{-1} which formerly, depending on one's point of view, presented either a paradox or an incontrovertible argument against continental drift, can now be explained as a fortuitous coincidence. Previous thinking took for its premise that heat flow was 1.5 and 1.6 μ cal.cm^{-2} sec^{-1} on land and at sea respectively from which McDonald (1965) calculated that at 35 km depth the heat flux was greater under the ocean by 0.4 to 1.2 μ cal.cm^{-2} sec^{-1}, implying deep-seated differences in temperature under continents and oceans. This gives some justification to statements by opponents of continental drift like: "At first we should notice that the idea of drift of continents is absolutely not compatible with the distribution of heat flow on the surface of the earth". In the same article Beloussov (1965) says: ". . . if we have the irresistible desire to move the continents, we have to move them together with the upper mantle. This signifies that the drift must involve not only a thin sheet of granitic layer as was originally supposed by Wegener, but a block 1,000 km thick".

CRUSTAL MODEL

The preferred geochemical model of the transition of continental shield to ocean compatible with modern data and current speculations can be much thinner than 1,000 km as shown on Fig.116. The bottom of the lithosphere is the 1,500°C isotherm which

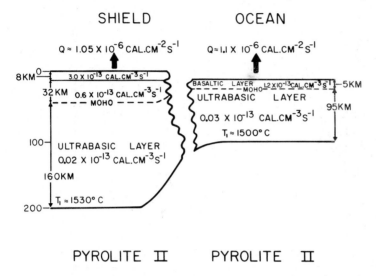

Fig.116. Geochemical model of continental shield and oceanic lithosphere (after Ringwood, 1969, fig.5) assuming convection maintains a constant temperature at the base of the two lithospheres. (From Sclater and Francheteau, 1970.)

might be regarded as the top of the low velocity layer and which, because of its inherent mobility is probably close to being isothermal. The equilibrium heat flow of about 1.1 μ cal.cm^{-2} sec^{-1} can be obtained by the choice of heat sources indicated in the figure. In all such models the radioactive elements have been concentrated upward by melting because the large atomic diameter of potassium, uranium and thorium prevents these elements from being easily incorporated into the crystal lattices of the silicates in the presence of a liquid phase. Other models can be devised, and the interested reader should consult the original papers.

CALCULATION OF OCEAN DEPTH FROM AGE

In calculating the deepening of the ocean as the rock column contracts due to cooling, one assumes that the ocean floor is isostatically compensated (Talwani et al., 1965), i.e., that rock columns of unit cross-section and 100 km deep everywhere weigh the same, and that it is only their density that changes with age. Heat flow and topography have been calculated as a function of distance from the East Pacific Rise (Fig.117) on the assump-

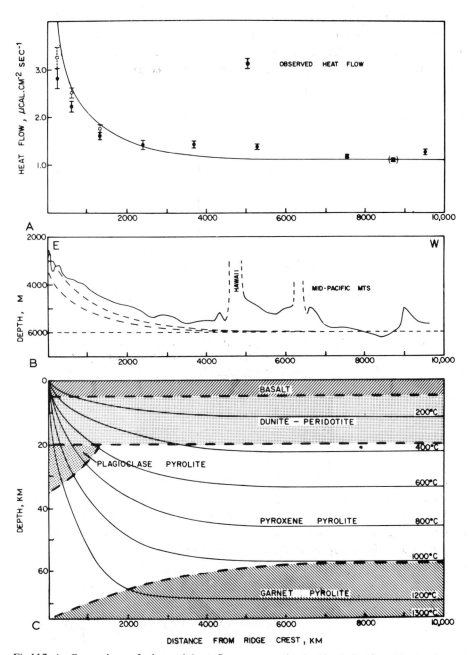

Fig.117. A. Comparison of observed heat flow averages in the North Pacific with the theoretical profile for a 75-km thick lithosphere. B. Comparison of observed topography (solid line) along 20°N with two theoretical profiles. The upper dashed curve is the profile expected from the thermal expansion and the phases of the model shown in (C). The lower dashed curve is the profile assuming thermal expansion of a lithosphere of uniform density. C. Isotherms and chemical zoning of a 75-km thick lithosphere moving at 5 cm/year to the right. (From Sclater and Francheteau, 1970.)

tion that material issues horizontally from the vertical plane containing the ridge axis at a
uniform initial temperature of 1,300°C. Comparison with observations shows curves of
similar shape which could be altered to produce a better fit by adjusting the many
parameters within acceptable limits. In Fig.118 depth is plotted against age or anomaly

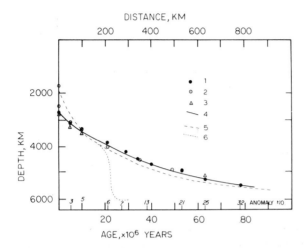

Fig.118. Plot of elevation of ridge versus age for the North Pacific, South Atlantic and Indian Oceans.
1 = North Pacific; *2* = South Atlantic; *3* = Indian Ocean; *4* = observed average. The theoretical profile
(*5*) is for a 100-km-thick lithosphere with a basal temperature of 1,475°C. The light dashed line (*6*)
represents the theoretical elevation across a ridge which started to spread 20 m.y.B.P. at the half-rate
of 1 cm/yr. (From Sclater and Harrison, 1971.)

number by averaging selected profiles in the Pacific, Atlantic and Indian Oceans. The
points are compared to the computed curve for a lithosphere 100 km thick with a basal
temperature of 1,475°C. By matching bathymetric profiles with the theoretical model
one can find spreading ridges where magnetic anomalies are too weak or too broken up to
be reliably identified (Sclater et al., 1971, in preparation). The southwest branch of the
Indian Ocean Ridge is such a case (Fig.119). Another application in regions younger than
60 m.y. is to look for topographic expression of ridges that have jumped or are just
starting to spread.

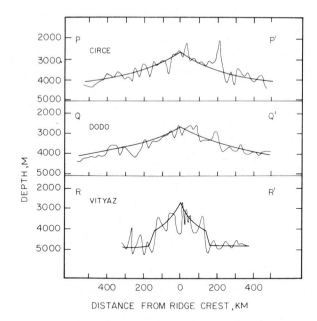

Fig.119. Topographic profiles across the central (*PP'*), southeast (*QQ'*) and southwest (*RR'*) branches of the Indian Ridge. See Fig.105 for location. For *PP'* and *QQ'* the calculated topography shown by the heavy smooth curve was obtained from the spreading rate of 1 cm/year measured by identifying anomalies 3 and 5. The same spreading rate was used to calculate the bathymetric profile *RR'* where anomalies could not be deciphered. From the good match between the measured and the computed depth profiles Sclater and Harrison (1971) concluded that the southwest branch is spreading at 1 cm/year. The sharp uplift 170 km from the axis of ridge marks start of present spreading episode (see Chapter 9). (From Sclater and Harrison, 1971.)

CHAPTER 11

THE NORTH ATLANTIC

Magnetic anomalies in the Atlantic Ocean confirm its origin by sea-floor spreading as postulated from the fitting of continental shores and slopes of the bordering continents by Wegener (1929), Carey (1958) and Bullard et al. (1965).

The opening of the Atlantic happened from the relative motion of the North American, Eurasian and African plates, creating different magnetic patterns north and south of the Azores. In general, anomalies are poorly developed and interpretations depend on recognizing a few most prominent ones.

FOUR MAGNETIC ZONES

Four distinct zones appear on Fig.120 which presents the magnetic remains of sea-floor spreading in the southwestern North Atlantic. Along the 28th parallel the standard geomagnetic reversal chronology can be followed from the ridge at 42°W to anomaly 32 (78 m.y.B.P.) at 55°W. Then follows an anomaly labeled R which can be recognized on profiles B-A1 and B-A2 near latitude 19°N. From 58°W to 67°W there is a region of confused anomalies of considerable amplitude which cannot be identified from one profile to the next. Between 67°W and 71.5°W an orderly set of lineated anomalies called the Keathley sequence extends from 23°N to 34°N. West of anomaly J20 we find the "quiet zone" which consists of magnetic anomalies of 5–20 γ amplitude. The transition from rough to smooth magnetic field occurs rather abruptly, within about 150 km.

A similar rough-smooth boundary is present off the northwestern coast of Africa (Fig.121) with the mirror image of the Keathley sequence to the east of it (Fig.122). Several hypothetical modes of origin for the quiet zone have been reviewed by Vogt et al. (1970a). In the Northeast Pacific the smooth zone seems to start right after anomaly 32, whereas in the North Atlantic and in the South Pacific anomaly 32 is followed by unidentified anomalies of large amplitude. It is therefore unlikely that the quiet zone arose from lack of magnetic reversals or from mineralogical alterations caused by rapid sedimentation at the spreading center but that it probably is the result east–west spreading in equatorial latitudes, as proposed by Vogt et al. (1970a). Shortly after the splitting of America from the Old World in the Triassic, the equator went through the quiet zone as determined from paleomagnetic measurements on rocks from Europe and North America. The poor precision of the data on which this more attractive explanation is based appears in the scatter of the paleomagnetic poles in Fig.123. The quiet zone boundary according to this interpretation does not represent an isochron.

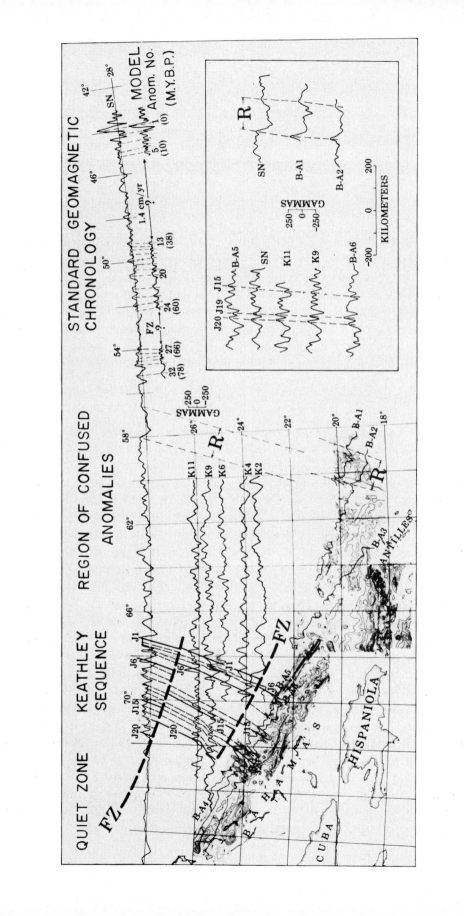

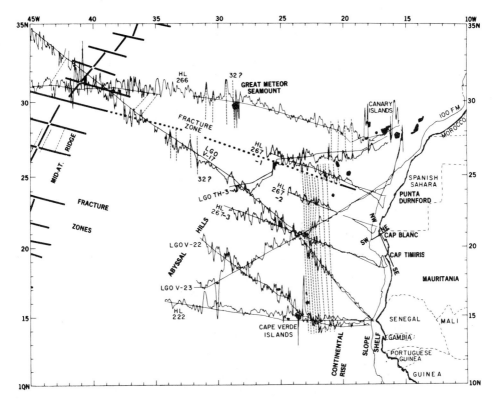

Fig.121. Residual magnetic profiles recorded by Hudson Laboratories (*HL; USNS* J.W. Gibbs) and by Lamont-Doherty Geological Observatory (*LGO*) are plotted along ship's tracks between the Canary and Cape Verde islands. Magnetic anomalies are tentatively correlated with dashed lines. Islands and seamounts are shaded. The axis of the Mid-Atlantic Ridge and fracture zones are shown (Heezen and Tharp, 1968, quoted in Rona et al., 1970). An apparent right-lateral offset in the north—south linear magnetic anomalies south of the Canary Islands approximately coincides with the dashed projection of the Atlantis fracture zone, which offsets the Mid-Atlantic Ridge a comparable direction and amount near 30°N. (Rona et al., 1970.)

Fig.120. Magnetic anomaly zones in the southeastern North Atlantic Ocean. Correlations between profiles of anomalies older than the standard chronology are presented in the inset. Along 28°N the boundary of the continental slope comes at 75°30'W. (From Vogt et al., 1971.)

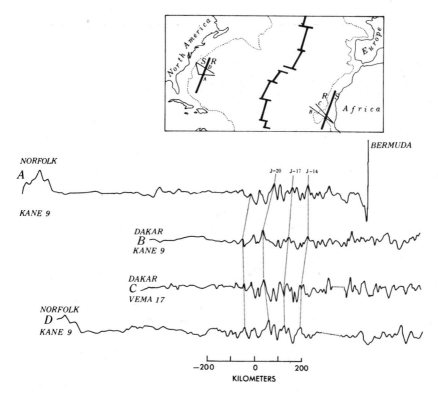

Fig.122. Rough-smooth magnetic boundary and several anomaly correlations between the eastern and western Atlantic (Schneider et al., 1969). Note that smooth-rough transition occupies several anomaly wavelengths. (Vogt et al., 1970a.)

VARIABLE SPREADING RATES

The spacing of the recognized anomalies gives the history of the spreading rates. In Fig.124 selected observed magnetic profiles were projected on lines bearing N104°E which is the average spreading direction south of the Azores. Model anomalies were calculated for comparison at 34.5°N and 26°N using the average measured spreading rates for the intervals between anomalies numbered according to the standard chronology. The spreading, rate indicated by numbers under the model profiles, was adjusted between the identified anomalies to fit best the data from many more profiles than are shown in the figure.

A similar example (Fig.125) was constructed for the area north of the Azores by projecting observed profiles on N88°E. The non-uniformity of the spreading rate is more pronounced, being only 0.4 cm/year between anomalies 13 and 5, a time interval of 29 m.y. To get the complete history of the opening of the North Atlantic one has to rely on ages from JOIDES holes and the physiographic fit of the continents along the 1,000-m

TRIASSIC

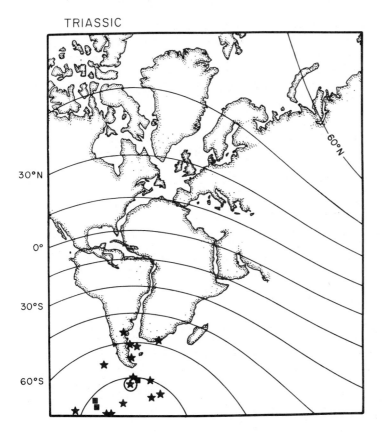

Fig.123. Paleogeography of the Atlantic continents in Triassic times (fit of Bullard et al., 1965). North America is kept fixed. The paleolatitudes inferred from the paleomagnetic poles of the four continents are drawn every 15°. Paleomagnetic south poles from Europe (stars) and North America (squares) appear at the bottom of the map. The mean pole from Europe is surrounded by a circle. Mercator projection. (Francheteau, 1970.)

isobath of Bullard et al. (1965) shown in Fig.123. Although the initial rifting might be timed by the age of the Palisades lava flows near the mouth of the Hudson River (190–200 m.y. ago, Erickson and Kulp, 1961), ages from drilling and spreading rates indicate that steady spreading probably began 180 m.y. ago. Table V compares the ages from JOIDES holes with the isochrons of Fig.126. The latter was constructed on the assumption that the shape of the ridge stayed the same, which is justified by the strike of the Keathley sequence, and the morphological fit of Bullard et al. (1965). In Fig.126 the key magnetic anomalies on the east side of the ridge were rotated to coincide with the corresponding anomalies on the west side about the centers listed in Table VI.

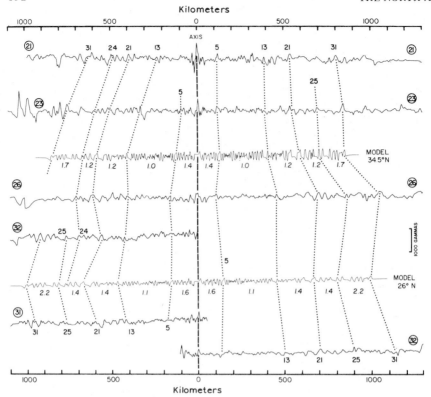

Fig. 124. The numbered profiles are projected along an azimuth of 104°, an average spreading direction for the Cenozoic south of the Azores–Gibraltar Ridge. The half spreading rates are indicated below the model profiles. (Pitman and Talwani, 1971.)

TABLE V

Comparison of ages obtained from JOIDES drill holes with ages from isochrons in Fig. 126[1]
(From Pitman and Talwani, 1971)

Site	Lat. N	Long. W	Depth (m)	Oldest sediment	Was basement reached?	Age	Age predicted from isochron map of Fig. 126
4	24 28.7	73 47.5	5319	Tithonian	no	140	156
9A	32 46.4	59 11.7	4965	Maestrichtian	no	70	103
10	32 51.7	52 12.9	4697	Middle and Lower Campanian	yes	80	75
11	29 56.6	44 44.8	3556	Middle Miocene	yes	18	17
12	19 41.7	26 00.0	4557	Eocene	no	51	137
99	23 41.1	73 51.0	4914	Oxfordian	no	155	164
100	23 41.3	73 48.0	5325	Callovian	yes	162	164
101	25 11.9	74 26.3	4868	Tithonian	no	140	162
105	34 53.7	69 10.4	5251	Oxfordian	yes	155	157

[1] Since this table was compiled hole #136 was drilled near 34°10'N 16°18'W at the rough-smooth magnetic boundary on the African side. It gave an age of only 110 m.y.B.P. (Anonymous, 1971) for the sediment immediately above the basalt. This single discordant datum should be considered spurious so long as it is not confirmed by additional evidence.

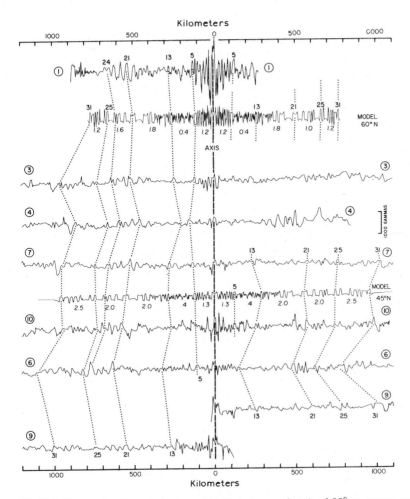

Fig. 125. The numbered profiles are projected along an azimuth of 88° an average spreading direction for the Cenozoic north of the Azores–Gibraltar Ridge. The half-spreading rates are indicated below the model profiles. (Pitman and Talwani, 1971.)

TABLE VI

Finite rotations required to rotate anomalies on the west and east side of the north Mid-Atlantic Ridge into coincidence, as shown on Fig.126. Asterisks denote pole position obtained from analysis of fracture zones
(From Pitman and Talwani, 1971)

Anomaly number	Age time scale of Heirtzler et al. (1968) (m.y.)	Latitude of pole of rotation	Longitude of pole of rotation	Rotation (degrees)
A. North of Azores				
5	9	56.3 N	141.4 E	2.24
13	38	58.0 N	147.0 E	4.55
21	53	48.0 N	153.0 E	10.07
25	63	63.0 N	157.0 E	14.00
31	72	81.0 N	167.0 E	21.52
B. South of Azores				
5	9	69.7 N	33.4 W	3.60
13	38	79.0 N	13.0 E	9.75
21	53	77.0 N	15.0 E	13.90
25	63	75.0 N	15.0 E	17.00
31	72	71.0 N	10.0 W	24.00

Fig.126. The open circles, triangles, diamonds and squares are locations of identified anomalies. The closed circles, triangles, diamonds and squares are locations obtained by rotating the anomaly locations on the east side of the ridge through the total rotations listed in Table VI. The rotations are, of course, different for each anomaly and different again south of the Azores–Gibraltar Ridge and north of it. The near coincidence of the closed circles, triangles, etc. demonstrates the validity of the rotations listed in Table VI. The dotted flow lines describe the motion of the continents on either side with respect to the ridge axis and their parallelism to the fracture zones also constitutes support for the validity of the derived rotations. The parallelism of the New England Seamount Chain and the Canary Islands is to the same flow line and suggests that these chains of islands constitute a fracture zone. Similarly the escarpment off the southern part of the Bahama platform lies parallel to a flow line and may thus be an old fracture zone. The 200-m and 1,000-m isobaths are shown. The dashed lines on which the anomaly symbols are located are labeled outward on both sides from the ridge axis up to 180 m.y.B.P. and represent isochrons which can be compared with ages of JOIDES deep-sea drilling shown by dots with numbers giving the age of the oldest rock reached (Ewing et al., 1969, 1970; Peterson et al., 1970). Eartquake epicenters are denoted by small dots and the boundary of the magnetic quiet zone is indicated by a wide dashed line. Crosses within circles are locations of buried diapiric structures. (From Pitman and Talwani, 1971.)

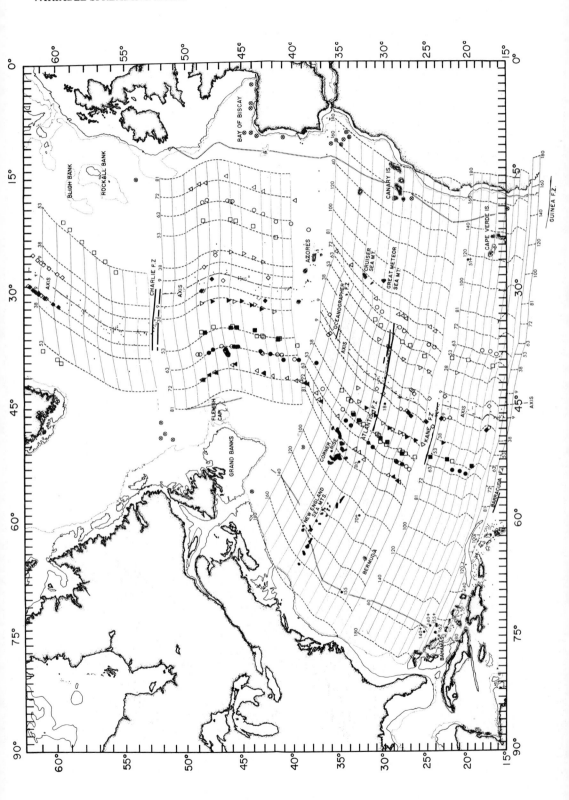

VARIATIONS IN SPREADING DIRECTION

The detailed survey of the area between the Reykjanes Ridge and Rockall Bank shows on Fig.127 a more complex history than what one might surmise from Fig.126. According to this survey, at anomaly 24 time (60 m.y. ago) the strike of the ridge was N42°E with a spreading rate of 0.75 cm/year. This situation lasted roughly until 45 m.y. ago producing an anomaly pattern which was nearly continuous. Then the axis of the ridge started to swing northward until about anomaly 10 time (32 m.y.B.P.) when its strike became almost north–south, causing many transform faults to break up the original linear pattern, as indicated by the trace of anomaly 13 (38 m.y.B.P.). By anomaly 6 time (21 m.y. ago), the ridge swung back to nearly its original orientation where we find it now, striking N36°E, and spreading at 1 cm/year. Previously we have seen a monotonic change in direction of spreading exhibited by the fracture zones and anomalies in the northeastern Pacific (Fig.97–99). In Fig.127 we observe a return to the original spreading direction. Such a pattern can be produced either by the migration of the center of relative rotation of the lithospheric plates or by the boundaries between the plates departing

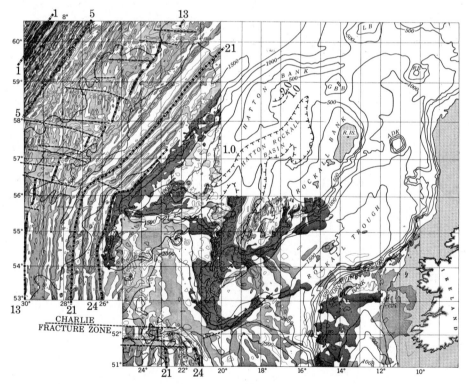

Fig.127. Magnetic anomaly chart between Reykjanes Ridge and Rockall Bank. The contour interval is 200 γ. The ridge crest at the northwestern corner of the chart is designated as *1*. Anomalies 5, 13, 21 and 24 are marked. Note the broken up pattern of spreading between anomalies 21 and 5. (Vogt et al. 1971.)

from the simple geometry in which the strike of the magnetic anomalies lies along great circles through the center of rotation and transform faults are arcs of small circles about that center. If it is the center of rotation that had shifted, contemporaneous shifts of relative motion involving adjoining lithospheric plates can be expected to be found in the geological records at the boundaries of these other plates. One could thus seek to relate the often easily dated record of plate rotations on the ocean floors to the geological history on the continents where plate rotations are difficult to discern.

ROTATION OF SPAIN

The small overlap of Spain on North Africa after rotation (Fig.123) would have been much larger had Spain not been rotated to close the Bay of Biscay, as was first done by Du Toit (1937) and then by Carey (1958). We have now factual evidence from two sources to confirm this guess, viz., a fan-shaped pattern of unidentified magnetic anomalies in the Bay of Biscay (Fig.128) and paleomagnetic measurements on rocks of the Iberian Peninsula which give the amount of rotation as well as some age limits for when it took place.

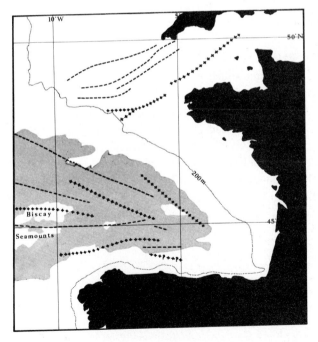

Fig.128. Trends of magnetic anomalies (+ and −) in the Bay of Biscay. The angle measured between the outer anomalies is $35° + 5°$. Shading indicates the flat sea floor of the Biscay Abyssal Plain and the lower part of the continental rise, which lie deeper than 2,400 fm (4,400 m). (Matthews and Williams, 1968.)

The magnetic anomalies in the Bay of Biscay exemplify formation of sea floor by rotation of a small plate situated near the center of rotation. The 5,000 m depth of the water, which is proportional to the age of the ocean floor (Sclater et al., 1971), suggests that rotation began at least 50 m.y. ago.

The paleomagnetic data reviewed by Girdler (1965) gave Spain a rotation of $38° \pm 10°$ relative to Europe since the Eocene. Van Dongen (1967) found $30°$ in the eastern Pyrenees, and Van der Voo (1967) measured $32°$ in central Spain after Triassic time. Watkins and Richardson (1968) got $22°$ since the Eocene on Lisbon volcanics, although their interpretation has been questioned (Van der Voo, 1969).

Because at anomaly 32 time the North Atlantic was nearly closed, and because the pole of rotation between North America and Europe was at $81°$N, one would expect the Cretaceous paleomagnetic pole of North America to be valid for Spain after rotating Spain to close the Bay of Biscay. Van der Voo (1969) has shown that to bring the Cretaceous paleomagnetic pole for Spain into coincidence with the North American pole, Spain has to be rotated clockwise $15°$ between the Late Cretaceous and the Oligocene, the counterclockwise rotation having been completed prior to the Late Cretaceous. This helps explain compressional geologic features in sedimentary strata in the Bay of Biscay and the west coast of Portugal as reviewed by Pitman and Talwani (1971).

GREENLAND AND THE LABRADOR SEA

The opening of the Norwegian Sea since anomaly 24 time (60 m.y. ago) indicates that this is the earliest date that Greenland became attached to the North American plate. This does not preclude some subsequent slow spreading in the Labrador Sea which would still fall within the uncertainty with which the spreading rate between the Eurasian and the North American plate can be measured. Pitman and Talwani (1971) identify the inactive Labrador Sea Ridge as anomaly 19 (47 m.y.B.P.). Vogt and Ostenso (1970) give it a slightly younger age: 42 m.y.B.P. Pitman and Talwani (1971) close the Labrador and the Norwegian Seas 81 m.y. ago. This is about 15 m.y. earlier than the date inferred by Francheteau (1970) from paleomagnetic data, but in view of the accuracy of such deductions these results corroborate rather than contradict each other. Thus, about 47 m.y. ago, the Labrador Sea acquired its present width and Greenland became attached to North America.

A SINGLE SUPERCONTINENT

In the preceding sections we have closed the North Atlantic 200 m.y. ago. The South Atlantic was already closed for some time before, resulting in the paleogeographic reconstruction pictured on Fig.123. But Antarctica, India and Australia can be fitted onto Africa resulting in the supercontinent called Gondwanaland (Fig.129) for the existence of which there is considerable paleomagnetic evidence from Triassic to Cambrian time, although the Triassic pole for Australia does not fit as well as the others. According to the magnetic anomaly record of sea-floor spreading, Australia separated from Antarctica only at anomaly 18 time (46 m.y.B.P.). The discrepancy can be caused either by a shuttling motion of Australia which failed to leave a geologic record or by errors in paleomagnetic data which according to Creer (1970) is unlikely. The polar path of the southern continents is shown by the heavy line in Fig.129 and proves that the supercontinent has moved as a unit with respect to the earth's spin axis, suggesting that plate motions existed 800 m.y. before it broke up to form the present continents. When North American and European poles are included (Fig.130) we are left with the impression that from the Cambrian to the Triassic all the land area of the world consisted of a single land mass called *Pangea* in the literature on continental drift. Paleomagnetic data do not exclude the possibility of relative motion of $10°-20°$ between parts of this supercontinent (Creer, 1970), a welcome reservation to those geologists who interpret the sequence of some pre-Triassic formations on the edges of the Precambrian nuclei as resulting from subduction at ancient island arcs (Dewey and Bird, 1970). For instance, to explain the Paleozoic

Fig.129. Paleomagnetic poles for the Gondwanic continents plotted against the background of Sproll and Dietz' morphological fit of Antarctica to Africa for pre-Triassic time. A common curve for all the continents is shown: erect triangles represent South American poles; inverted triangles represent African poles; squares represent Australian poles; diamonds represent Indian poles and dots represent Antarctic poles. (Creer, 1970.)

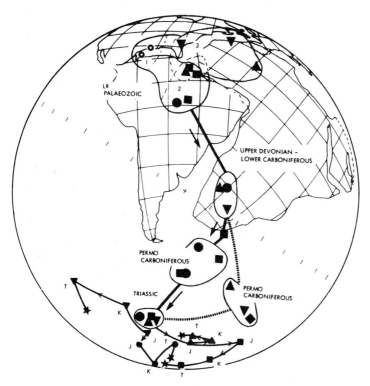

Fig.130. Paleomagnetic poles for the four Atlantic continents: North America, Europe, South America and Africa. Poles represented by squares, dots, erect and inverted triangles, respectively. A common curve represented by the thicker black line can be drawn through these poles from the Triassic back in time. The Mesozoic and Tertiary drift phase is represented by the diverging PW curves (thin black lines) for post-Triassic time. Two Permo-Carboniferous groups of poles are shown, one for Laurasia and the other for Gondwanaland (note the Australian pole is included here and represented by the diamond). It is not established yet whether this divergence of the Laurasian and Gondwanaland curves is real; the latter is shown as a broken line. The Lower Palaeozoic group of poles has been sub-divided into areas *1*, *2* and 3. The group which plots in southern Africa possibly dates from the Silurian rather than from the Upper Devonian (as labelled). (Creer, 1970.)

geology of eastern North America on this basis, a proto-Atlantic Ocean is needed. These relative motions, however, if they did exist were non-dispersive and many times smaller than the drift of Pangea as a whole, which is confirmed by the grouping of continental nuclei which are more than 1,700 m.y. old in the reconstruction of Pangea in Fig.131. For an unknown reason, the distribution of the major occurrences of granulite such as eclogite facies rocks, shown as black dots in the figure, exhibits an even more remarkable coherence.

There is thus good evidence that prior to 200 m.y. ago all the land on our globe drifted as a unit which then broke up into the continents as we know them. The uniqueness of this event might be the starting point of speculations on the mechanism responsible for plate motions.

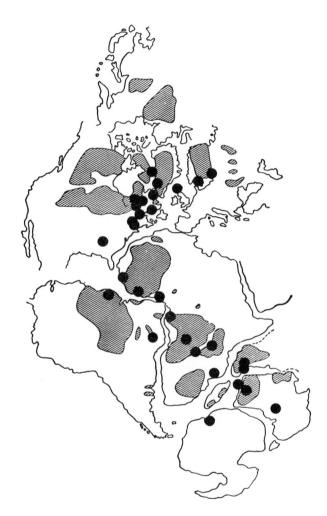

Fig.131. Some of the major occurrences of deep-seated facies rocks plotted as black circles along with the groupings of Precambrian rocks older than 1700 m.y. shown hashured on the reconstruction of Pangea. The coherence of this distribution is unexplained, but again suggests no drift prior to the drift of 200 m.y. ago. (Hurley and Rand, 1969. Copyright by the American Association for the Advancement of Science.)

CHAPTER 12

THE ARCTIC OCEAN

FROM ICELAND TO NANSEN RIDGE

North of Iceland the Mid-Atlantic Ridge connects with the Nansen Ridge of the Arctic Ocean via intermediate ridge segments and fracture zones in the Norwegian Sea outlined by earthquake epicenters and bathymetry (Fig.132). The Nansen Ridge abuts against the Siberian continental shelf about 78°N 130°E. From there, the epicenters become sparse and diffuse and one has to guess that the northern boundary between the Eurasian and American plates ends via the Verkhoiansk mountains in a triple junction with the Pacific plate near the base of Kamchatka.

Lineated magnetic anomalies are present parallel to the Iceland—Jan Mayen Ridge and to the Mohns Ridge. Some ridges and fracture zones are identified by number in Fig.132. The outstanding characteristic of this area is that ridge segments are not always parallel to each other or perpendicular to the fracture zones. The magnetic profiles across the Iceland—Jan Mayen Ridge are shown on Fig.133. Profile 4 in Fig.133 correlates reasonably well with the standard model.

Avery et al. (1968) show a tentative identification of anomalies up to No.24 (60 m.y.B.P.) southeast from the Mohns Ridge (Fig.134). If this identification is correct, it implies that the spreading rate between anomalies 24 and 20 was three times faster than between anomaly 20 and the ridge crest. South of the Jan Mayen Fracture Zone another set of lineated anomalies occurs between the Jan Mayen Ridge and Norway (Fig.135). From their symmetry and the width of the Norwegian Basin, Vogt et al. (1970b) proposed that the pattern was created by an earlier episode of spreading, which ended about 46 m.y.B.P. corresponding to anomaly 18 time. Much later, spreading was resumed from the Iceland—Jan Mayen Ridge. Further north, the Mohns Ridge remained in its original position, which accounts for its being half way between the Greenland and the Norwegian shelves. Thus the Norwegian Sea is an example of episodic or non-uniform sea-floor spreading.

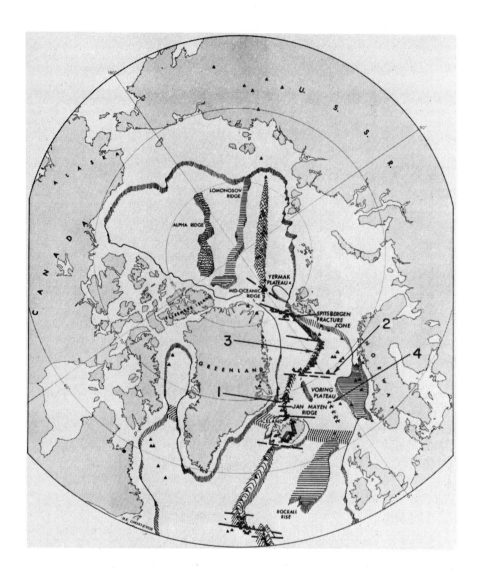

Fig.132. Schematic chart of physiographic features in deep-water areas of Arctic. Triangles denote earthquake epicenters. Symbols denoting buried ridges and partially emergent ridges are self-explanatory (from Johnson, 1969). *1* = Iceland–Jan Mayen Ridge; *2* = Jan Mayen Fracture Zone; *3* = Mohns Ridge; *4* = Inactive ridge. (From Vogt et al., 1970b.)

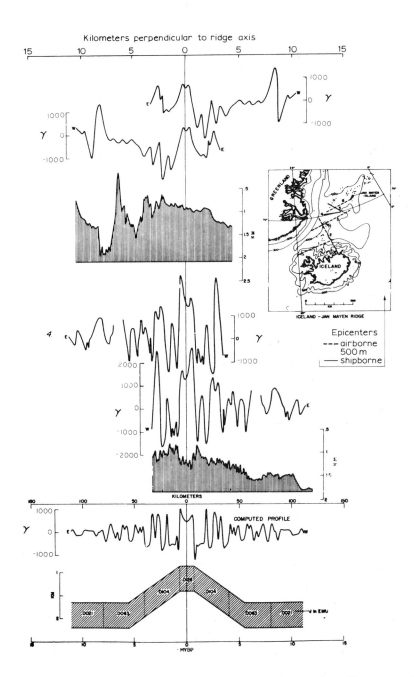

Fig.133. Shipborne magnetic anomalies, bathymetry and magnetic profiles computed from standard reversal chronology of Heirtzler (1968) for model spreading at 1 cm/year across Iceland–Jan Mayen Ridge. All data projected perpendicular to ridge axis. (From Vogt et al., 1970b.)

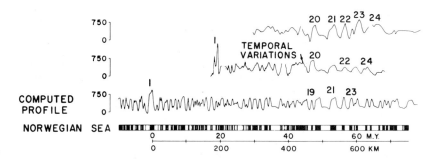

COMPUTED
PROFILE

NORWEGIAN SEA

Fig.134. Comparison of observed anomalies across the Mohns Ridge with a profile based on the model of Heirtzler et al. (1968). The Mohns Ridge is designated by *1* on Fig.132. (After Avery et al., 1968.)

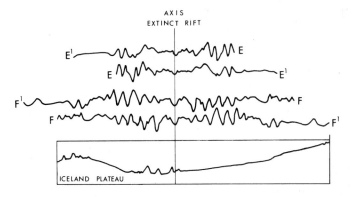

Fig.135. Reinterpretation of reversed pairs of profiles EE^1 and FF^1 (Avery et al., 1968, fig.2) in terms of extinct ridge hypothesis. Ridge is designated by *4* on Fig.132. (From Vogt et al., 1970b.)

THE NANSEN RIDGE

The aeromagnetic map of the Nansen Ridge in Fig.136 shows that the central anomaly consists of four segments, segment *CD* being roughly parallel to segment *AB*, while segments *BC* and *DE* although not parallel to each other are oriented clockwise with respect to the other two. The segment *BC* is roughly parallel to the spreading direction of the Reykjanes Ridge for the last 10 m.y. The apparent orientation of the ridge does not reveal the actual direction of spreading, since the ridge must be interrupted by transform faults which are not visible on the magnetic chart, as we have seen, for instance, in the case of the Indian Ocean Ridge. If, however, we assume that the Nansen Ridge anomaly pattern resulted from the relative rotation of the Eurasian and North American plates, we can use the dated spreading of the Reykjanes Ridge to establish its rate of spreading. The double spreading rate from the Reykjanes Ridge back to 60 m.y. is about 2 cm/year. The total width of the lineated magnetic anomaly pattern of the Nansen Ridge is about 5.5°

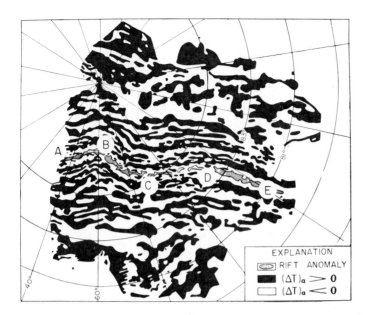

Fig.136. Aeromagnetic anomaly chart of the portion of the Arctic Ocean surrounding the Nansen Ridge (*ABCDE*). The anomalies cannot be identified by the standard chronology. If the spreading resulted from relative rotation of the Eurasian and North American plates about the centers of Table VI during the last 60 m.y. as indicated by the Reykjanes Ridge (Fig.127), the spreading rate should be 0.55 cm/year, which is consistent with the approximate width of 5.5° (610 km) of the lineated pattern. (Trubiatchinskii et al., 1970.)

or 610 km. Dividing this distance by 60 m.y. yields 1.1 cm/year. The ratio of 2 to 1.1 should equal the ratio of the cosines of the distances of the Reykjanes and Nansen ridges from the center of rotation of the North American and Eurasian plates, which happens to be the case, these distances being 30° and 63°, respectively. We may conclude that spreading of the Nansen Ridge has been going on for the last 60 m.y. at the slow rate of 0.55 cm/year. Magnetic anomalies associated with the Nansen Ridge have defied identific-ation with the standard chronology, although Vogt et al. (1970b) have attempted to identify them in a profile of unknown location published by Rassokho et al. (1967) shown in Fig.137. The computed profile has many more features than the observed one, even at nearly twice the spreading rate.

When the Eurasian plate is rotated clockwise about the centers in Kamchatka to close the spreading from the Nansen Ridge, the aseismic Lomonosov Ridge fits nicely into the continental shelf of Siberia. Pitman and Talwani (1971) interpret the Lomonosov Ridge (Fig.132) as having been a part of the Siberian shelf which was split off the continent by spreading of the Nansen Ridge (Fig.138) which began 63 m.y. ago (anomaly 25) and lasted to the present day. The formation of the Norwegian Sea would be contempo-raneous with the spreading of the Nansen Ridge. Geological evidence reviewed by Pitman and Talwani (1971) suggests that prior to the Late Cretaceous the relative motion of the

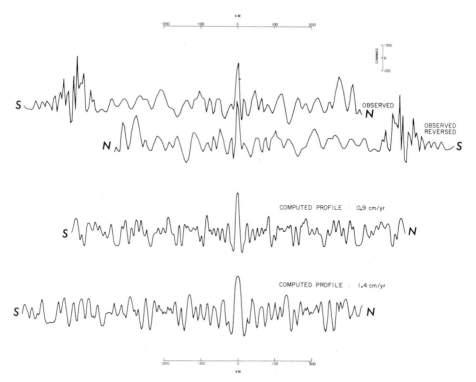

Fig.137. Aeromagnetic residual profile across Nansen Ridge (Rassokho et al., 1967), compared with two spreading models, suggests a rate less than 0.9 cm/year in the axial region. Model magnetized layer is 5 km below observation level, 1.7 km thick, and magnetized at ±0.0022 e.m.u. except ±0.01 for axial block. (From Vogt et al., 1970b.)

Eurasian and North American plates caused compression in the Arctic and in the Bering Sea. The inactive Alpha Ridge (Fig.132), however, is flanked by magnetic anomalies (Ostenso, 1962) which have not been identified but which suggest a sea-floor spreading origin. From this evidence Vogt and Ostenso (1970) have proposed that spreading took place from the Alpha Cordillera from 60 to 40 m.y. ago and that the present spreading from the Nansen Ridge began 40 m.y.B.P. This does not preclude compression in the Arctic prior to 60 m.y. ago. However, if the Alpha Cordillera was formed by sea-floor spreading prior to either 40 or 60 m.y. ago, it should have had time to cool down and its present relief of 2 km cannot be accounted for by the simple mechanism of Chapter 10. The difficulty can be solved by having the Alpha and the Nansen ridges opening up simultaneously or several times in quick succession with the Alpha Ridge being quiescent at the present time. We would have then a small independent plate between the Alpha and the Nansen ridges whose other boundaries might perhaps be defined by future surveys.

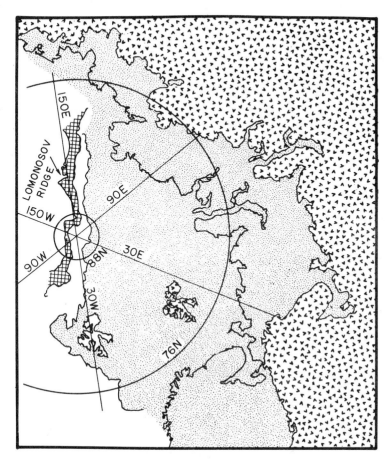

Fig.138. Physiographic fit of the Lomonosov Ridge and Siberian continental shelf by rotation about the centers of Table VI. The Nansen Ridge is presumed to have split the Siberian continental shelf of which the Lomonosov Ridge is the western remnant. (Pitman and Talwani, 1971.)

CHAPTER 13

MAGNETIC INTENSITY MEASUREMENTS NEAR AND ON THE BOTTOM

Magnetic measurements at depth in the ocean are of two kinds. One consists of towing a total intensity magnetometer near the bottom for investigating the fine structure of the magnetic features. The other one simultaneously records time variations in three components of the magnetic intensity at several locations on the bottom of the ocean for exploring the distribution of electrical conductivity aand therefore of the temperature in the mantle rock.

NEAR BOTTOM MAGNETIC SURVEYS

In addition to the proton precession magnetometer, the "Deep Tow" instrument package towed by the ship at about 2 knots, carries acoustic and photographic equipment listed and illustrated in Fig.139. Power, data and command signals are transmitted from

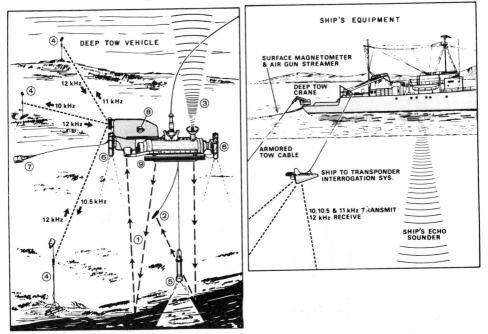

Fig.139. The 'Deep Tow' instrument and the towing ship showing full instrument operation. *1* = Downward looking echo sounder; *2* = 3.5 kHz 'P' bottom penetration; *3* = upward looking echo sounder; *4* = acoustic transponder; *5* = strobe light; *6* = camera; *7* = magnetometer; *8* = sound velocimeter; *9* = side-looking sonar. (Spiess and Mudie, 1970.)

and to the towing ship by a single conductor co-axial armored tow cable 10 km long. Usually three or four acoustic transponders are planted on the bottom preferably near the summit of topographic elevations as determined by the ship's echo sounder. The position of the tow is obtained with a precision of 5–20 m from acoustic ranging on these beacons. Areal surveys measure on the order of 15 nautical miles (n.m.) on the side, and profiles about 100 n.m. long have been observed. Fig.140 shows one of these long profiles run

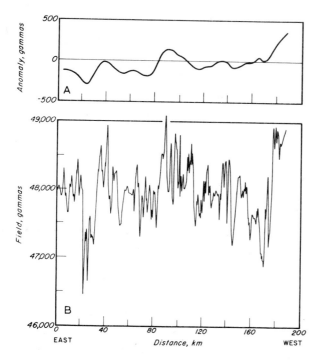

Fig.140. East-west magnetometer profile near 34°N 127°W. A. On the surface of the ocean. B. At an average distance of 40 m from the bottom. (Spiess and Mudie, 1970.)

east–west across the magnetic pattern in the Northeast Pacific near 34°N 127°W. Sharp anomalies of 1,000 γ amplitude and a wavelength of 3–4 km are superposed on the broader anomalies. The latter are the only ones one measures at the sea surface. The same can be said of the profiles in Fig.141 surveyed 9° to the east of the profile of Fig.140, and which show coherence of features between profiles. From this and other examples, Spiess and Mudie (1970) conclude that the orientation of the small scale magnetic and topography features tends to run parallel to the lineated magnetic anomalies observed at the sea surface. The relation between topography and magnetic anomalies should be clearer at ridge crests where sedimentation does not mask the topography. However, profiles across the East Pacific Rise at the mouth of the Gulf of California (Fig.142) and the Gorda Ridge (Fig.143) show that there is no apparent coherence between bottom topography and near-bottom magnetic anomalies.

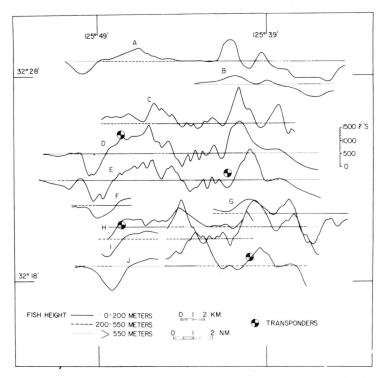

Fig.141. East–west magnetic profiles taken with Deep Tow instrument showing coherence at anomalies between profiles. The height of the instrument above bottom is indicated by symbols. (Spiess and Mudie, 1970.)

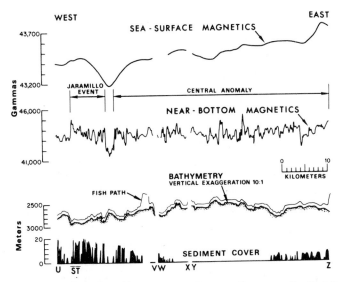

Fig.142. Profile of near-bottom magnetic anomalies across the East Pacific Rise crest at the mouth of the Gulf of California at 21°N 109°W (Fig.100). Note the random character of the near-bottom magnetic trace and the absence of sediment cover of the spreading center. (Larson et al., 1968.)

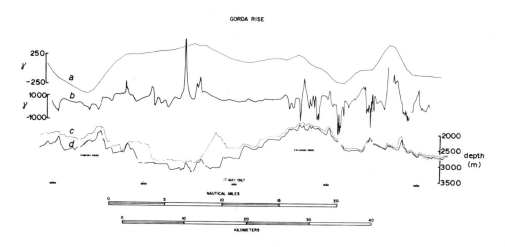

Fig.143. Magnetic profile across the Gorda Rise in the Northeast Pacific near 41°N 127°W. a = Magnetic intensity at the surface; b = magnetic intensity near the bottom; c = depth of Deep Tow; d = bathymetric profile. (Spiess and Mudie, 1970.)

The amplitude and the sharpness of magnetic anomalies near the bottom prove that the upper surface of the magnetic rock layer comes close to the surface of the bottom and that it probably consists of many individual lava flows which contribute to the smooth anomalous magnetic intensity at the surface of the ocean. It is tempting to interpret some of the short period magnetic anomalies observed near the bottom in terms of short-time variations in the strength and direction of the geomagnetic dipole field, but until such time as identical anomaly sequences are found in different parts of the world, the contribution of the variations of the dipole field cannot be separated from local causes. It is unlikely that there shall be sufficient data of this kind to refine the geomagnetic reversal time scale because the method is slow and relatively expensive compared to paleomagnetic measurements on marine sediments.

MAGNETIC SHORT-TIME VARIATIONS ON THE OCEAN FLOOR

The electrical conductivity of rocks rises sharply by two orders of magnitude at 920°C (Akimoto and Fujisawa, 1965). The rise of the geotherms as one approaches a spreading ridge (Fig.144) constitutes a lateral change of electrical conductivity, which can be represented by a horizontal conductor added on to a uniform conducting shell from which no anomaly can be expected. When the component of the earth's horizontal magnetic field which is perpendicular to this extra conductor changes in magnitude, an electric current is induced in the conductor in such a direction as to oppose the change in the magnitude of the inducing field. To the right and to the left of the horizontal conductor,

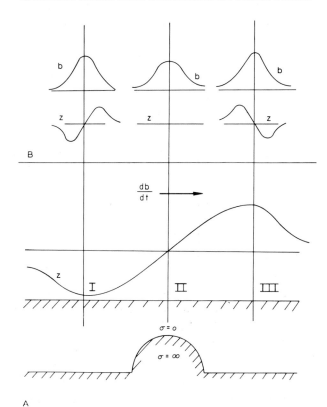

Fig.144. Detection of local rise of geotherms by geomagnetic time variations. A. The infinitely conducting region which might represent a spreading ocean ridge is modeled as a half-cylinder of infinite
length in the direction normal to the paper and separated from the surface of the ground by cold rock
of nearly zero conductivity. The component of the horizontal magnetic intensity b perpendicular to
the ridge axis changes at the rate db/dt, thus inducing an electric current in the semi-cylindrical
conductor in a direction such as to oppose db/dt. The vertical magnetic intensity produced by this
induced current is positive left of the conductor, zero directly over it and negative to the right. B. The
vertical field variation is shown for the three positions (*I-III*) in response to a horizontal field variation
b of period of 2 hours. Note that the phase of z is opposite on the two sides of the conductor. The
horizontal field of the induced current decreases the observed horizontal field b at location *II* more
than at location *I* and *III*.

the induced current produces vertical magnetic variations of opposite sign, as shown
diagramatically in Fig.144A. A number of such anomalies comprehensibly reviewed by
Rikitake (1971) have been found in continental areas, as for example the one in Fig.145.
At the continental boundaries, the conductivity of the ocean water produces a strong
anomaly in the magnetic time variations called the "coast effect" which render observatories situated near the coasts of large oceans unsuitable for correcting ocean magnetic
surveys for time variations. The presence of an anomalous horizontal linear conductor is
best revealed by the ratio of the vertical variation to that component of the variation in
the horizontal plane which is most coherent with the vertical variation. The ratio is
plotted as a vector parallel to this horizontal component. It is often called the "Parkinson

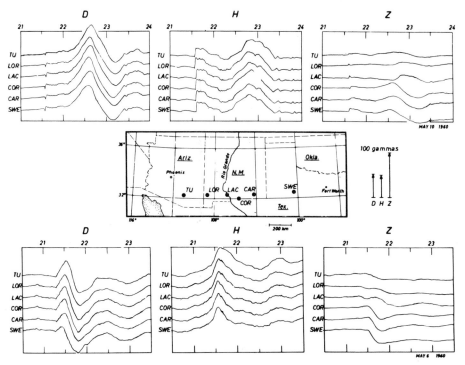

Fig.145. Two bay events as recorded on the profile Tucson-Sweetwater through southern Arizona, New Mexico and West Texas. Distinct increase of Z-amplitude east of the Rio Grande, preceded by a reversal between Las Cruces and Cornudas. (From Schmucker, 1970.)

(1962) vector". It points toward the good conductor, in the case of the coast effect, toward the ocean. The calculation is carried out or periods from 15 min. to one day. The effect of the shallowing of the geotherms from continent to ocean is in the same direction as the coast effect making it difficult to separate one from the other. Bullard and Parker (1970) have developed a method for calculating magnetic effects due to a realistic model of the world ocean for the first few harmonics of the diurnal variation, so as to get the contribution from the conductivity of the mangle rock by subtracting the calculated variations from the observed ones. In Peru, Casaverde et al. (1967) and Richards (1970) have found small Parkinson vectors at the coast pointing landward, suggesting elevated geotherms under the Andes which more than balance out the conductivity of water and mantle of the Pacific. At present, such local differences in the coast effect contribute the only reliable indication that the electrical conductivity and therefore the temperature of the mantle may not change everywhere in the same way across the continental boundaries.

Temporary magnetic observatories on the ocean bottom have been built and operated experimentally since 1965 off the California coast (Cox et al., 1970). They permitted to extend Schmucker's (1970) and White's (in preparation) geomagnetic deep sounding surveys into the ocean. A record of a small polar storm taken on the ocean bottom is presented in Fig.29. Parkinson vectors shown on Fig.146 appear normal at the ocean

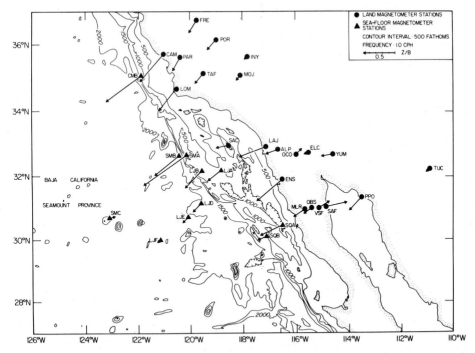

Fig.146. Parkinson vectors in and off the coast of southwestern North America for geomagnetic time variations of 1-hour period. The length of the arrows is proportional to the ratio of the amplitude of the vertical variation to the amplitude of that component of the horizontal variation which is most coherent with the vertical variation. The arrow is parallel to the horizontal variation and points toward the good conductor. A segment of the East Pacific Rise appears to lie at the head of the Gulf of California. The coast effect dies out rapidly off the coast, which is probably helped by the concentration of electric current flow at the edge of the ocean. (Greenhouse, 1971; White, 1972.)

boundary, but indicate conducting material at the head of the Gulf of California. This material can not be the shallow water or basin sediments because the latter are insulated from both the ocean and the conducting mantle by cold non-conducting rock. The conduction is ascribed to a segment of the spreading ridge electrically connected to the hot and therefore conducting mantle material.

The ocean-bottom temporary magnetic observatories can but have not yet been used along with the measurement of electric fields in the ocean (Filloux, 1967a) for magneto-telluric sounding of the assumed horizontally stratified conductivity structure which should exist far from land and other anomalous conditions in the hope of establishing a value for the depth to the asthenosphere and its temperature. In his magneto-telluric experiment Filloux (1967b) used an ingenious single-component horizontal magnetometer which consists of a magnet suspended on a fiber, and an optical system with a photoelectric sensor. At present, there have not been a sufficient number of geomagnetic and magneto-telluric soundings at sea to warrant even tentative interpretations in terms of mantle temperatures. Although the tools for making the measurements have been developed, their use is slow and expensive enough to require another 5 or 10 years for the gathering of a significant number of data.

CHAPTER 14

POSSIBLE FUTURE DEVELOPMENTS

In broad outline, sea-floor spreading with plate theory as its consequence may be regarded as permanent achievements of the new global geology. Magnetic anomalies have been mapped in the key areas of the World Ocean. From them and from paleomagnetic data paleogeographic reconstructions of the continents have been made back to Triassic time (200 m.y.B.P.). We may look for remnants of older ocean floor which have escaped from being swallowed up by the trenches. We may expect more efforts to relate the Keathley sequence of Jurassic age in the Atlantic to the undated magnetic anomalies in the Western Pacific (Hayes and Pitman, 1971; Vogt et al., 1971). To date, such attempts are not convincing. Additional paleomagnetic surveys of seamounts with more emphasis on their age determination would help in areas where the sea floor is older than the Late Cretaceous.

The world-wide termination of the correlatable magnetic anomalies 80 m.y. ago and their resumption in the Keathley sequence 130 m.y.B.P. should be confirmed by additional magnetic surveys at sea and perhaps also by more refined paleomagnetic measurements of reversals in JOIDES cores. Since it is now possible to re-enter the hole, drilling can continue to the crystalline basement rocks. A meaningful reversal chronology will depend on how completely cores are recovered.

The festooning of island arcs with convex curvature toward the subducting plate and other geomorphologic and petrologic evidence (Karig, 1970, 1971) suggests that some marginal seas expand, pushing the islands and the trench associated with them toward the ocean. In Fig.65 lava produced by friction of the subducting plate feeds the andesitic volcanoes of the island arc, but in addition if it is to supply the high heat flow in the Seas of Okhotsk and Japan (Yasui et al., 1968a,b) vast quantities of molten basalt must rise into the bottom of these seas (Hasebe et al., 1970), for which there is no room unless the seas are spreading. Such spreading may be maintaining the western end of the Aleutian Trench where it is parallel to the motion of the Pacific plate, and also the Puerto Rico Trench which runs parallel to the spreading direction in the Atlantic. Magnetic anomalies in the Sea of Japan do not help to clarify this general problem, but magnetic surveys in the South Fiji Basin (Sclater et al., 1971) do indicate anomalies characteristic of sea-floor spreading between Tonga and the Lau Ridge. A similar situation is suspected in the Philippine Sea, where there is geological evidence for a spreading center on the concave side of the island arc associated with fast subduction (Karig, 1971). Perhaps some information on the nature of the forces responsible for sea-floor spreading and subduction might come from detailed magnetic surveys of such complex areas.

The lack of contradictory experimental facts is a cause for the lack of progress we may expect in the future development of plate theory. It is unlikely that additional magnetic measurements at sea will materially change its main features, although they will provide details of the motions of small plates associated with short spreading centers like the Galapagos Rift located at 1°N between 86° and 92°W (Raff, 1968).

A re-examination of continental geology from the point of view of plate theory might be fruitful (Dewey and Horsfield, 1970; Dewey and Bird, 1971), but not too much help should be expected from magnetic anomalies on land which are complex compared to the lineated ones at sea and do not lend themselves to comparable unique interpretation.

Future contributions of geomagnetism to unraveling paleogeography is likely to come from paleomagnetic investigations on the continents and also in a minor way in the oceans.

INTERNATIONAL GEOMAGNETIC REFERENCE FIELD 1965.0
(I.A.G.A. Commission 2, 1969)

To obtain the magnetic anomaly the smoothed geomagnetic field called the Reference
Field needs to be subtracted from the observed values. To take care of secular variations
the reference field has to be corrected to some standard date called "epoch", like 1965.0.
The earth's internal geomagnetic field is given in geocentric spherical coordinates: θ, the
colatitude measured from the North Pole, and λ the longitude reckoned eastward from
Greenwich. The radius a of the reference sphere representing the earth is 6371.2 km. The
magnetic potential of the earth is expressed as a series of spherical harmonics:

$$V = a \sum_{n=1}^{8} \sum_{m=0}^{n} \left(\frac{a}{r}\right)^{n+1} (g_n^m \cos m\lambda + h_n^m \sin m\lambda) P_n^m (\cos \theta)$$

where r is the distance from the center of the reference sphere to a point where the field
is sought which should lie on or above the earth's surface. The numbers g_n^m and h_n^m are
the Schmidt quasi-normalized harmonic coefficients (Chapman and Bartels, 1940,
pp.610–612) of which there are 80 for $n = m = 8$. The function $P_n^m (\cos \theta)$ is an
associated Legendre function of degree n and order m which, setting $\mu = \cos \theta$, can be
written:

$$P_n^m (\mu) = \frac{1}{2^n n!} \left(\frac{\epsilon_m (n-m)! (1-\mu^2)^m}{(n+m)!}\right)^{1/2} \cdot \frac{d^{m+n} (\mu^2-1)^n}{d\mu^{m+n}}$$

where $\epsilon_m = 1$ for $m = o$, and $\epsilon_m = 2$ when $m \geqslant 1$. The components of magnetic intensity
are:

$$X = \frac{1}{r} \frac{\partial V}{\partial \theta}, \quad Y = -\frac{1}{r \sin \theta} \frac{\partial V}{\partial \lambda}, \quad Z = \frac{\partial V}{\partial r}$$

The scalar intensity T (or F), which is the quantity measured by the total intensity
magnetometer is

$$T = (X^2 + Y^2 + Z^2)^{1/2}$$

If the value of harmonic coefficient is $g_n^m(t_o)$ at the reference epoch $t_o = 1965$, its
value at another time t is

$$g_n^m(t) = g_n^m(t_o) + \dot{g}_n^m(t - t_o)$$

The table lists the harmonic coefficients and their rates of change for the epoch 1965. Grid values of the total magnetic intensity every two degrees have been tabulated by Fabiano and Peddie (1969) along with a Fortran program for calculating the geomagnetic elements and their rates of change.

CALCULATION OF TOTAL MAGNETIC INTENSITY PROFILE PERPENDICULAR TO A RIDGE MODEL

A modern treatment of the calculation of total magnetic intensity of a two-dimensional strip model adapted to digital computers was published by Talwani and Heirtzler (1964). Most of the model anomalies for sea-floor spreading shown in this book were essentially calculated by this method. The model of the ocean floor consists of infinite strips parallel to the ridge and magnetized either parallel or antiparallel to the dipole part of the geomagnetic field at the latitude. Usually the thickness and magnetization of the strip are held constant. Sometimes the thickness and the magnetization vary with distance to better fit the observed profiles, but in every case it is assumed that the magnetic dip angle and the strike of the whole pattern have remained the same to the present day. As long as the spreading direction was east—west these assumptions were acceptable, but in the Indian Ocean where the ancient spreading was north—south, McKenzie and Sclater (1971) had to take into account changes in latitude and azimuth between the epochs of generation and observation of the anomalies. Even then the

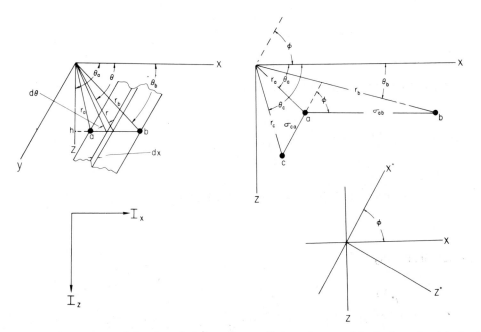

Fig.147. Geometry used for the calculation of magnetic intensity caused by horizontal and inclined magnetized planes infinite in the *y* direction.

improved system still contains simplifying assumptions, viz., that the latitude and azimuth of the ridge generating all the visible anomalies did not vary, but that between the birth of the pattern and the observation, both latitude and orientation changed. The present exposition was constructed from an interpretation of a computer program written by D.P. McKenzie. A condensed version of the method constitutes appendix 2 of McKenzie and Sclater (1971). The comparison between observed and computed profiles is one of trial and error. Profiles as they would be observed at the present latitude and orientation are computed from the standard field reversal chronology and assumed spreading rates for several original latitudes and orientations which might be the likely ones as guessed from other information. These profiles are then compared to the observed profiles perpendicular to the strike of the anomalies. The ones that "appear" to fit best are chosen as evidence for the paleogeographic reconstruction.

In Fig.147 the magnetometer is at the origin with the y axis parallel to the strike of the strip model which extends to infinity in both directions. We shall first calculate the horizontal and vertical magnetic intensity at the origin due to the upper surface of the magnetized strip at the depth h and whose width between points a and b is $x_a - x_b$. In keeping with our elementary point of view, we can consider the magnetization of the top surface of a normally magnetized strip to be represented by a uniform density of south poles which we have likened to negative masses. The attraction at the origin of the mass element $\sigma_{ab}\ dx\ dy$ is $\dfrac{\sigma_{ab}\ dx\ dy}{r^2}$, and for the infinite line element it is:

$$\int_{-\infty}^{\infty} \frac{\sigma_{ab}}{r^2}\ dx\ dy = \frac{2\sigma_{ab}}{r}\ dx$$

The force in the x direction from the surface (ab):

$$\Delta H_x\ (ab) = \int_{x_a}^{x_b} \frac{2\sigma_{ab}}{r} \cos\theta\ dx = 2\sigma_{ab} \int_{\theta_a}^{\theta_b} \frac{\cos\theta}{r} \frac{r\,d\theta}{\sin\theta}$$

$$= 2\sigma_{ab} \int_{\theta_a}^{\theta_b} \cot\theta\ d\theta = \log\sin\theta \ \bigg|_{\theta_a}^{\theta_b} = 2\sigma_{ab} \log\frac{r_a}{r_b} \tag{1}$$

The vertical intensity:

$$\Delta Z\ (ab) = \int_{x_a}^{x_b} \frac{2\sigma_{ab}}{r} \sin\theta\ dx = 2\sigma_{ab} \int_{\theta_a}^{\theta_b} \frac{\sin\theta}{r} \cdot \frac{r}{\sin\theta}\ d\theta = 2\sigma_{ab}\ (\theta_b - \theta_a) \tag{2}$$

The log in eq.2 is negative and so is $(\theta_b - \theta_a)$ but σ_{ab} is also negative because it is populated by south poles, so both results are positive.

Next we calculate the contribution from the surface (ca) of Fig.147, which is also

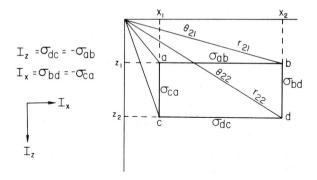

$$I_z = \sigma_{dc} = -\sigma_{ab}$$

$$I_x = \sigma_{bd} = -\sigma_{ca}$$

Fig.148. The rectangle $abcd$ represents the section of a uniformly magnetized parallelopiped infitely long in the y direction.

infinite in the $\pm y$ directions and which makes an angle with the horizontal. For this, we rotate the x and z axes about the y axis by the angle so as to make the rotated x-axis parallel to the face (ca). We can write eq.1 and eq.2 with respect to the rotated coordinate system x^* and z^*:

$$\Delta H_{x*} = 2\sigma_{ca} \log \frac{r_c}{r_a}$$

and:

$$\Delta Z^* = 2\sigma_{ca} (\theta_a^* - \theta_c^*)$$

The components in the original coordinates can be written:

$$\Delta H_x (ca) = \Delta H_{x*} \cos \phi + \Delta Z^* \sin \phi$$

and:

$$\Delta Z (ca) = \Delta Z^* \cos \phi - \Delta H_{x*} \sin \phi$$

Substituting the values of $\Delta H_x{}^*$ and ΔZ^* and noting that $\theta_a^* - \theta_c^* = \theta_a - \theta_c$, we get:

$$\Delta H_x (ca) = 2\sigma [\cos \phi \log \frac{r_c}{r_a} + (\theta_a - \theta_c) \sin \phi]$$

and:

$$\Delta Z (ca) = 2\sigma_{ca} [\sin \phi \log \frac{r_c}{r_a} - (\theta_a - \theta_c) \cos \phi]$$

If the face (ca) is vertical, as is the case of the sea-floor spreading models, $\phi = \pi/2$ and the preceding equations degenerate into:

$$\Delta H_x (ca) = 2\sigma_{ca} (\theta_a - \theta_c) \tag{3}$$

and:

$$\Delta Z (ca) = 2\sigma_{ca} \log\frac{r_c}{r_a} \tag{4}$$

In Fig.148, the rectangle *abcd* represents a vertical section through one of the magnetized strips of the model. Previously we had calculated the contributions to the magnetic intensity at the origin of the faces *ab* and *ca* of the strip. The contributions of the other two faces can be written by inspection and added after changing sign because opposite faces have opposite polarity. The ficticious surface distributions can also be replaced by the realistic volume magnetization intensities $I_x = -\sigma_{ca}$ and $I_z = -\sigma_{ab}$. The complete expressions for the rectangular strip become:

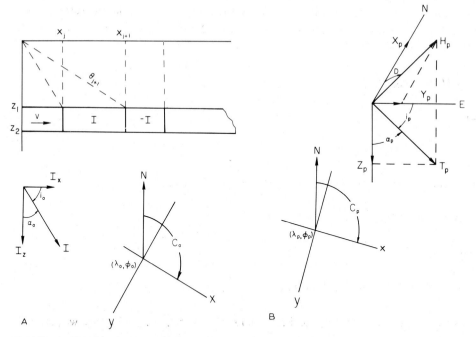

Fig.149. A. The ridge axis y spreads at uniform speed v, with the spreading direction x making an angle C_0 with magnetic north N. I_x is the component of horizontal polarization parallel to the direction of spreading x and i_0 the dip angle at the geographic coordinates $(\lambda_0\phi_0)$ at the time of spreading. B. At present the ridge is at position (λ_p, ϕ_p) and the spreading direction is C_p while the dip angle is now i_p. The value T_p of the I.G.R.F. and its components are shown in the upper right-hand part of the figure.

$$\Delta H_x \, (abcd) = 2I_x \, (\theta_{21} + \theta_{12} - \theta_{11} - \theta_{22}) + 2I_z \left(\log \frac{r_{21}}{r_{11}} - \log \frac{r_{22}}{r_{12}}\right) \qquad (5)$$

and:

$$\Delta Z \, (abcd) = 2I_x \left(\log \frac{r_{12}}{r_{11}} - \log \frac{r_{22}}{r_{21}}\right) + 2I_z \, (\theta_{11} + \theta_{22} - \theta_{21} - \theta_{12}) \qquad (6)$$

in which the r's and the θ's are indexed with their coordinates x and z, running between n and $-n$ for x and z having the values z_1 or z_2. In setting up the computation it is convenient to replace the angles and the distances in terms of x and z, as for example:

$$\theta_{21} = \tan^{-1} \frac{z_1}{x_2}, \text{ and } 2 \log \frac{r_{22}}{r_{21}} = \log r_{22}^2 - \log r_{21}^2$$

where:

$$r_{22}^2 = x_2^2 + z_2^2 \text{ and } r_{21}^2 = x_2^2 + z_1^2, \text{ etc.}$$

The complete model consists of a continuous string of rectangles such as $abcd$ of Fig.148 of thickness $z_2 - z_1$ with vertical boundaries at x_j which are calculated from the chronology of geomagnetic field reversals (Fig.41) at times t_j and the spreading velocity v so that the width of each block is $x_{j+1} - x_j = v(t_{j+1} - t_j)$ (Fig. 149A). Obviously, the model need not be restricted to just two values of z. The depth to the top and bottom may be different for each rectangle.

The components of magnetization depend on the strength M_0 of the dipole moment of the earth and on the latitude λ_0 of the place with respect to the pole at the time the pattern had formed. The horizontal component I_x also depends on the strike C_0 of the ridge. The magnetic dip angle i_0 at the time of formation is obtained from the latitude λ_0:

$$i_0 = \frac{1}{2} \tan^{-1} (90° - \lambda_0),$$

while the total intensity:

$$T_0 = \frac{M_0}{R^3} \sqrt{1 + 3 \cos^2 (90° - \lambda_0)}$$

both of these expressions being direct consequences of the formula for the field of a dipole. Taking $M_0 = 8.09 \cdot 10^{25}$, which is its present value, the ratio $M_0/R^3 = 0.32 \cdot 10^5$ γ. The total magnetization $I = T_0 k$ where k is the apparent susceptibility. The magnetization constants inserted into eq.5 and eq.6 are

$$I_x = I \cos i_0 \cos C_0 \text{ and } I_z = I \sin i_0$$

The contributions of all the blocks are then added to obtain ΔH_x and ΔZ at the origin. For other points on the profile one calculates the sums after displacing the pattern by small increments like 2 km.

At the present geographic position (λ_p, ϕ_p) the ridge makes an angle C_p with geographic north. At the same place the International Geomagnetic Reference Field (I.G.R.F.) $T_p (\lambda_p, \phi_p)$ has the components shown on Fig.149B. The horizontal components of the reference field normal and parallel to the strike of the anomalies are:

$$H_{xp} = T_p \cos i_p \cos (C_p - D_p)$$

and:

$$H_{yp} = T_p \cos i_p \sin (C_p - D_p)$$

The vertical component:

$$Z_p = T_p \sin i_p$$

The total magnetic intensity T is the sum of T_p and the contribution from the magnetized model:

$$T = [(H_{xp} + \Delta H_x)^2 + H_{yp}^2 + (Z_p + \Delta Z)^2]^{1/2} \tag{7}$$

To obtain the anomalous field we have to subtract the reference field from eq.7:

$$\Delta T = T - T_p \text{ (I.G.R.F.)}$$

This results is exact, i.e., we did not have to assume that the anomaly is small compared to the total field.

When no displacement between the times of formation and observation has occurred, $C_0 = C_p$ and T_0 can be taken as the present value of the dipole field for the location. Although we know from paleomagnetic evidence that the earth's magnetic moment has not remained constant through geologic time, its variations are within the uncertainty of the apparent susceptibility k. Since the intensity of magnetization is equal to the product of these two quantities, its value is adjusted to best fit the amplitudes of the observed profiles.

MAGNETIC PROPERTIES OF OCEANIC BASALTS

(From Jean Francheteau, personal communication, 1970)

Sample location	Depth	Specimen description	Intensity of natural remanent magnetization[1] ($J_n \cdot 10^{-3}$ e.m.u. cm^{-3})	Volume magnetic susceptibility (10^{-3} (k) cm^{-3})	Apparent susceptibility[2] (k_{app}) (10^{-3} cm^{-3})	Q-ratio[3]	Reference	Notes
Northeast Pacific Ocean 28°59'N 112°30'W	3738 m	Depth below top of basalt flow in cm					Cox and Doell (1962)	Average
		24	10.10	0.20	21.20	105		$J_n = 5.38 \cdot 10^{-3}$
		32	5.65	0.20	12.00	59		e.m.u./cm^3
		66	6.46	0.22	13.64	61		$(S.D. = 1.88 \cdot 10^{-3})$
		71	7.25	0.20	15.40	76		$k = 0.31 \cdot 10^{-3}$
		151	4.61	0.35	9.80	27		$(S.D. = 0.19 \cdot 10^{-3})$
		155	4.44	0.23	9.43	40		$Q = 40$
		184	6.10	0.36	12.96	35		$(S.D. = 20)$
		189	8.70	0.43	18.49	42		$T_{curie} = 355°C$
		222	5.96	0.25	12.75	50		titaniferous
		287	6.03	0.36	12.96	35		magnetite in
		322	5.19	0.37	11.10	29		skeletal grains
		334	5.70	0.23	12.19	52		$H_0 = 0.48$ Oe
		404	4.90	0.25	10.50	41		mean inclination
		458	4.60	0.27	9.72	35		of magnetization
		473	4.47	0.25	9.50	37		$= -36°$
		480	3.89	0.22	8.36	37		
		490	3.38	0.26	7.28	27		age 42 m.y.
		531	4.03	0.29	8.70	29		(Kulp, 1963)
		537	4.63	0.28	10.08	35		
		1250	4.89	0.32	10.56	32		
		1267	8.41	1.17	18.72	15		
		1275	2.21	0.27	4.86	17		
		1284	2.20	0.20	4.80	23		

For footnotes, see p. 173

Sample location	Depth	Specimen description	Intensity of natural remanent magnetization[1] ($J_n \cdot 10^{-3}$ e.m.u. cm⁻³)	Volume magnetic susceptibility (10^{-3} (k) cm⁻³)	Apparent susceptibility[2] (k_{app}) (10^{-3} cm⁻³)	Q-ratio[3]	Reference	Notes
Mid-Atlantic Ridge 30°01'N 42°04'W to 31°49'N 43°25'W		olivine basalt fresh	36.56	0.78	85.80	109	Opdyke and Hekinian (1967)	Q-ratio computed assuming $H_O = 0.43$ Oe average $k = 0.92 \cdot 10^{-3}$ absence of viscous remanent magnetization component
			13.55	0.86	32.68	37		
			11.64	1.06	28.62	26		
			15.91	0.65	38.35	58		
			13.28	0.85	32.30	37		
			26.11	0.83	62.25	74		
			22.19	0.34	53.58	56		
			22.68	0.67	53.60	79		
			38.96	0.56	90.16	160		
			24.99	0.50	58.50	116		
			33.77	0.74	78.44	105		
			30.55	0.38	71.82	188		
			16.13	0.24	36.96	153		
			10.74	0.50	25.50	51		
		altered	3.37	0.17	7.99	46		
		fresh	20.06	1.11	47.73	42		
			15.59	0.41	36.49	88		
			27.84	0.78	66.30	84		
		sphenitic	32.11	0.50	75.00	149		
		olivine free	3.81	5.14	15.42	2		
		sphenitic decomposed	1.89	0.14	4.76	33		
			3.31	2.89	8.67	2		
		altered	10.43	0.90	26.10	28		
		fresh	14.16	0.46	33.58	72		
Mid-Atlantic Ridge 45°11'N 27°56'W to 27°56'W	1440– 1350 fm	fine-grained basalts	1.69	0.63	4.41	6	Vogt and Ostenso (1966)	average $J_n = 5 \cdot 10^{-3}$ e.m.u./cm⁻³ $k = 0.3 \cdot 10^{-3}$ $Q = 48$
			7.29	0.26	15.34	58		
			7.79	0.26	16.90	64		
			11.50	0.40	24.40	60		
			9.40	0.16	19.52	121		
			4.62	0.87	10.44	11		

Location	Depth	Rock type					Reference	Remarks
44°34'N 28°09'W to 44°36'N 28°07'W	1740 fm	glass sheathed basalt fragments	6.80	0.15	14.55	96	Irving (1968)	used magnetic viscosity to infer polarity of young basalt
			2.32	0.29	5.22	17		
			0.76	0.06	1.74	28		
			0.56	0.20	1.40	6		
Mid-Atlantic Ridge 45°10'N		olivine tholeiite (pillow lava)	54.1					
North Pacific Cobb Seamount 46°46'N 130°43'W		fresh tholeiite (non oxidized)	1.77	0.22	2.42	10	Dymond and Windom (1968)	a density of 2.8 g cm^{-3} was assumed exceptional magnetic susceptibility in a.c. field demagnetization age: 1.6 m.y.
		altered-tholeiite (oxidized)	3.11	0.37	5.92	15		
Mid-Atlantic Ridge 22°30'N 45°30'W	3100–2530 m	olivine tholeiite medium-grained	3.63	1.57	9.42	5	Luyendyk and Melson (1967)	titanomagnetite locally oxidized to ilmenohematite
			3.60	1.88	9.40	4		
	3380–2495 m	coarse-grained	4.74	2.69	10.76	3		
			4.67	2.86	11.44	3		
			5.01	2.98	11.92	3		
	3525–3000 m	fine-grained	11.29	0.81	23.49	28		
			8.26	0.75	17.25	22		
			9.55	0.64	19.84	30		
	1985–1735 m	brecciated basalts	0.25	0.03	0.54	5		in breccias titanomagnetite replaced by goethite ilmenohematite sphene
			0.40	0.11	0.88	7		
			0.64	0.32	1.60	4		
			0.36	0.22	0.88	3		
			0.05	0.08	0.16	1		

For footnotes, see p. 173

Sample location	Depth	Specimen description	Intensity of natural remanent magnetization[1] ($J_n \cdot 10^{-3}$ e.m.u. cm^{-3})	Volume magnetic susceptibility (10^{-3} (k) cm^{-3})	Apparent susceptibility[2] (k_{app}) (10^{-3} cm^{-3})	Q-ratio[3]	Reference	Notes
		basalt fragment in breccia	0.09	0.09	0.27	2		
			0.13	0.13	0.39	2		
			2.23	1.58	6.32	3		
			2.45	1.85	7.40	3		
North Atlantic seamount 34°51'N 16°31'W	750 fm	vesicular olivine basalt	5.0	0.29	12.47	42	Langton et al. (1960)	Q-ratio computed by assuming $H_0 = 0.42$ Oe
			5.1	0.43	12.47	28		
			4.8	0.28	11.48	40		
			4.8	0.40	11.60	28		
			5.3	0.37	12.58	33		
North Pacific seamounts								
38°N 145°58'E	3000 m	basalt (1)	4.8	0.21	10.08	47	Ozima et al. (1968)	Tcurie range from 25°–600°C
37°03'N 163°45'E	3000 m	basalt (2)	1.3	0.20	2.80	13		
28°22'N 148°14'E	2000 m	alkali (3) olivine basalt	6.7	0.29	13.63	46		
27°03'N 148°33'E	1000 m	tholeiitic (4) basalt	4.3	1.1	9.90	8		3 out of 7 basalts showed self-reversal of TRM upon heating to 300°C in air
27°57'N 147°37'E	2000 m	basalt (5)	1.6	0.075	3.30	43		
20°45'N 112°47'W	712 m	olivine (6) basalt tholeiite	0.83	0.33	1.98	5		ages of samples: 72 m.y. (1) 25 m.y. (2) 64-74-79 m.y. (3) 27-96 m.y. (4) 18 m.y. (5) 2.4-2.8-3.3 m.y. (6)
21°7'N 119°22'N	2985 m	basalt	2.5	0.087	5.046	57		

Location	Depth	Rock type				Reference	Remarks
North Atlantic 41°21′N 14°28′W	4960 m	altered vesicular basalts				Matthews (1961)	histogram of apparent susceptibilities in 43 specimens of lava Median value $k = 0.5 \cdot 10^{-3}$ $J_n = 5 \cdot 10^{-3}$ e.m.u. cm^{-3}
Pacific and Atlantic Ocean	40–5370 m	90 basalts 2 gabbros				Ade-Hall (1964)	Pacific (18 sites) Atlantic (9 sites) 35 basalt samples are decomposed histograms of J_n, k, Q in 92 samples 3 classes of basalt defined by thermal demagnetization
North Pacific Mendocino Fracture Zone		basalts				Bullard and Mason (1963)	histogram of apparent susceptibilities in unknown number of samples median value $k_{app} = 35 \cdot 10^{-3}$ $Q = 20$
Average (number)			10.01 (78)	0.66 (77)	21.76 (77)	48 (77)	
Standard deviation			10.22	0.81	21.13	39	
Standard error			1.16	0.09	2.41	4	

[1] Intensity of magnetization, J_n is natural remanent magnetization (NRM).

[2] Apparent susceptibility $k_{app} = \dfrac{J_n + kH_O}{H_O}$ where H_O is the earth's field at the sampling location.

[3] Koenigsberger ratio $Q = \dfrac{J_n}{kH_O}$ where H_O is the earth's field at the sampling location.

REFERENCES

Ade-Hall, J.M., 1963. A correlation between remanent magnetism and petrological and chemical properties of Tertiary basalt lavas from Mull, Scotland. *Geophys. J.*, 8:403–423.

Ade-Hall, J.M., 1964. The magnetic properties of some submarine oceanic lavas. *Geophys. J.*, 9:85–92.

Akimoto, S. and Fujisawa, H., 1965. Demonstration of electrical conductivity jump produced by the olivine-spinel transition. *J. Geophys. Res.*, 70:443–449.

Anonymous, 1971. Deep sea drilling project, leg 14. *Geotimes*, 16(2):14–20.

Atwater, T., 1970. Implications of plate tectonics for the Cenozoic tectonic evolution of western North America. *Geol. Soc. Am. Bull.*, 81:3513–3536.

Atwater, T. and Menard, H.W., 1970. Magnetic lineations in the Northeast Pacific. *Earth Planet. Sci. Lett.*, 7:445–450.

Ave'Lallemant, H.G. and Carter, N.L., 1970. Syntectonic recrystallization of olivine and modes of flow in the upper mantle. *Geol. Soc. Am. Bull.*, 81:2203–2220.

Avery, O.E., Burton, G.D, and Heirtzler, J.R., 1968. An aeromagnetic survey of the Norwegian Sea. *J. Geophys. Res.*, 73:4583–4600.

Barazangi, M. and Dorman, J., 1969. World seismicity maps compiled from E.S.S.A., Coast and Geodetic Survey, epicenter data, 1961–1967. *Bull. Seismol. Soc. Am.*, 59:369–380.

Beloussov, V.V., 1965. The relationship between the Earth's crust and the deeper layers of the Earth. In: *Seminar on Earth Sciences, Hyderabad, India, 1964, Proc., Pt. I: Geophysics. Indian Geophys. Union Bull.*, 2:1–6.

Benioff, H., 1954. Orogenesis and deep crustal structure, additional evidence from seismology. *Geol. Soc. Am. Bull.*, 65:385–400.

Bolshakov, A.C. and Solodovnikov, G.M., 1969. Determination of the intensity of the ancient geomagnetic field from the magnetization of heated formations (by lava flows). In: G.N. Petrova (Editor), *Magnetism of Geological Formations and Paleomagnetism.* Institute of Physics of the Earth, Acad. Sci., U.S.S.R., Moscow, pp.129–131.

Bullard, E.C., 1967. The removal of trend from magnetic surveys. *Earth Planet. Sci. Lett.*, 2:293–300.

Bullard, E.C., 1968. The Bakerian Lecture 1967: Reversals of the Earth's magnetic field. *Philos. Trans. R. Soc. Lond., Ser. A*, 263:481–524.

Bullard, E.C. and Mason, R.G., 1961. The magnetic field astern of a ship. *Deep-Sea Res.*, 8:20–27.

Bullard, E.C. and Parker, R.L., 1970. Electromagnetic induction in the oceans. In: A.E. Maxwell (Editor), *The Sea.* Wiley, New York, N.Y., 4:695–730.

Bullard, E.C., Everett, J.E. and Smith, A.G., 1965. The fit of the continents around the Atlantic. *Philos. Trans. R. Soc. Lond., Ser. A*, 258:41–51.

Burlatskaia, S.P., Nachasova, I.E., Nechaeva, T.B. and Petrova, G.N., 1969. Archeomagnetic investigations. In: G.N. Petrova (Editor), *Magnetism of Rocks and Paleomagnetism.* Institute of Physics of the Earth, Acad. Sci., U.S.S.R., pp.166–169.

Cain, J.C., 1971. Geomagnetic models from satellite surveys. *Rev. Geophys. Space Phys.*, 9:259–273.

Carey, S.W., 1958. The tectonic approach to continental drift. *Symp. Cont. Drift, Hobart*, pp.177–355.

Carter, N.L. and Ave'Lallemant, H.G., 1970. High-temperature flow of dunite and peridotite. *Geol. Soc. Am. Bull.*, 81:2181–2202.

Casaverde, M., Giesecke Jr., A.A., Salgueiro, R., Del Pozo, S., Tamayo, L., Tuve, M.A. and Aldrich, L.T., 1967. Studies of conductivity anomalies under the Andes. *Carnegie Inst. Wash., Yearbook*, 66:369–372.

Chapman, S. and Bartels, J., 1940. *Geomagnetism.* Oxford Univ. Press, London, 1096 pp.

Chase, R.L. and Bunce, E.T., 1969. Underthrusting of the eastern margin of the Antilles by the floor of the western North Atlantic Ocean, and origin of the Barbados ridge. *J. Geophys. Res.*, 74:1415–1420.

Clegg, J.A., Almond, M. and Stubbs, P.H.S., 1954. The remanent magnetism of some sedimentary rocks in Britain. *Philos. Mag.*, 45:583–598.

Collinson, D.W., Creer, K.M. and Runcorn, S.K. (Editors), 1967. *Methods in Paleomagnetism*. Elsevier, Amsterdam, 609 pp.

Corry, C.E., Dubois, C. and Vacquier, V., 1968. Instrument for measuring terrestrial heat flow through the ocean floor. *J. Mar. Res.*, 26:165–177.

Cox, A., 1968. Length of geomagnetic polarity intervals. *J. Geophys. Res.*, 73:3247–3260.

Cox, A., 1969. Geomagnetic reversals. *Science*, 163:237–245.

Cox, A. and Doell, R.R., 1960. Review of paleomagnetism. *Bull. Geol. Soc. Am.*, 71:645–768.

Cox, A. and Doell, R.R., 1962. Magnetic properties of the basalt in hole EM 7, Mohole Project. *J. Geophys. Res.*, 67:3997–4004.

Cox, A., Doell, R.R. and Dalrymple, G.B., 1964. Reversals of the Earth's magnetic field. *Science*, 144:1537.

Cox, C.S., Filloux, J.H. and Larsen, J.C., 1970. Electromagnetic studies of ocean currents and electrical conductivity below the ocean floor. In: A.E. Maxwell (Editor), *The Sea*. Wiley, New York, N.Y., 4:637–693.

Creer, K.M., 1970. A review of palaeomagnetism. *Earth-Sci. Rev.*, 6:369–466.

Dewey, J.F. and Bird, J.M., 1970. Mountain belts and the new global tectonics. *J. Geophys. Res.*, 75:2625–2647.

Dewey, J.F. and Bird, J.M., 1971. Origin and emplacement of the ophiolite suite: Appalachian ophiolites in Newfoundland. *J. Geophys. Res.*, 76:3179–3206.

Dewey, J.F. and Horsfield, B., 1970. Plate tectonics, orogeny and continental growth. *Nature*, 225:521–525.

Dickson, G.O., Pitman III, W.C. and Heirtzler, J.R., 1968. Magnetic anomalies in the South Atlantic and sea-floor spreading. *J. Geophys. Res.*, 73:2087–2100.

Dietz, R.S., 1962. Ocean-basin evolution by sea-floor spreading. In: S.K. Runcorn (Editor), *Continental Drift*. Academic Press, New York, N.Y., pp.289–298.

DuToit, A.L., 1937. *Our Wandering Continents. An Hypothesis of Continental Drifting*. Oliver and Boyd, Edinburgh, 366 pp.

Dymond, J. and Windom, H.L., 1968. Cretaceous K-Ar ages from Pacific Ocean seamounts. *Earth Planet. Sci. Lett.*, 4:47–52.

Elsasser, W.M., 1955. Hydromagnetism I. A review. *Am. J. Phys.*, 23:590–609.

Elvers, D.J., Mathewson, C.C., Kohler, R.E. and Moses, R.L., 1967. *Systematic Ocean Surveys by the U.S.C. and G.S.S. „Pioneer" 1961–1963*. U.S. Coast Geodet. Surv., Dep. Commer., Operational Data Rep. C&GSDR-1: 19 pp.

Erickson, B.H. and Grim, P.J., 1969. Profiles of magnetic anomalies south of the Aleutian Island Arc. *Geol. Soc. Am. Bull.*, 80:1387–1390.

Erickson, G.P. and Kulp, J.L., 1961. Potassium-argon dates on basaltic rocks. *Ann. N.Y. Acad. Sci.*, 91:321–323.

Ewing, J., Hollister, C., Hathaway, J., Paulus, F., Lancelot, Y., Habib, D., Poag, C.W., Luterbacher, H.P., Worstell, P. and Wilcoxon, J.A., 1970. Deep sea drilling project. *Geotimes*, 15:14–16.

Ewing, M., Houtz, R. and Ewing, J., 1969. South Pacific sediment distribution. *J. Geophys. Res.*, 74:2477–2493.

Fabiano, E.B. and Peddie, N.W., 1969. *Grid Values of Total Magnetic Intensity IGRF-1965*. U.S. Dept. Commer., Rockville, Md. E.S.S.A. Tech. Rep. C&GS-38.

Filloux, J.H., 1967a. *Oceanic Electric Currents, Geomagnetic Variations and the Deep Electrical Conductivity Structure of the Ocean-Continental Transition of Central California*. Ph.D. Thesis, Univ. California, San Diego, Calif., 166 pp.

Filloux, J.H., 1967b. An ocean bottom, D component magnetometer. *Geophysics*, 32:978–987.

Fisher, R.L., Sclater, J.G. and McKenzie, D.P., 1971. The evolution of the Central Indian Ridge, Western Indian Ocean. *Geol. Soc. Am. Bull.*, 82:553–562.

Foster, J.H. and Opdyke, N.D., 1970. Upper Miocene to Recent magnetic stratigraphy in deep-sea sediments. *J. Geophys. Res.*, 75:4465–4473.

Francheteau, J., 1970. *Paleomagnetism and Plate Tectonics.* Ph.D. Thesis, Univ. California, San Diego, Calif.

Francheteau, J. and Sclater, J.G., 1969. Paleomagnetism of the southern continents and plate tectonics. *Earth Planet. Sci. Lett.*, 6:93–106.

Francheteau, J., Harrison, C.G.A., Sclater, J.G. and Richards, M.L., 1970. Magnetization of Pacific seamounts: a preliminary polar curve for the northeastern Pacific. *J. Geophys. Res.*, 75:2035–2061.

Girdler, R.W., 1965. Continental drift and the rotation of Spain. *Nature*, 207:396–398.

Grant, F.S. and West, G.F., 1965. *Interpretation Theory in Applied Geophysics.* McGraw-Hill, New York, N.Y., 583 pp.

Greenhouse, J.P., 1972. *Temporal Geomagnetic Field Variations on the Ocean Floor off Southern California.* Ph.D. Thesis, Univ. California, San Diego, Calif.

Griffiths, D.H., King, R.F. and Rees, A.I., 1962. The relevance of magnetic measurements on some fine-grained silts to the study of their depositional process. *Sedimentology*, 1:134–144.

Griggs, D., 1939. A theory of mountain building. *Am. J. Sci.*, 237:611–650.

Grim, P. and Erickson, B.H., 1969. Fracture zones and magnetic anomalies south of the Aleutian Trench. *J. Geophys. Res.*, 74:1488–1494.

Grow, J.A. and Atwater, T., 1970. Mid-Tertiary tectonic transition in the Aleutian Arc. *Geol. Soc. Am. Bull.*, 81:3715–3722.

Harrison, C.G.A., 1966. The paleomagnetism of deep-sea sediments. *J. Geophys. Res.*, 71:3033–3043.

Harrison, C.G.A., 1968. Formation of magnetic anomaly patterns by dyke injection. *J. Geophys. Res.*, 73:2137–2142.

Hasebe, K., Fujii, N. and Uyeda, S., 1970. Thermal processes under island arcs. *Tectonophysics*, 10:335–355.

Hayes, D.E. and Pitman III, W.C., 1971. Magnetic lineations in the North Pacific. *Geol. Soc. Am., Mem.*, 126:291–314.

Heezen, B.C., 1962. The deep-sea floor. In: K. Runcorn (Editor), *Continental Drift.* Academic Press, New York, N.Y., pp.235–286.

Heirtzler, J.R., Dickson, G.O., Herron, E.M., Pitman III, W.C. and Le Pichon, X., 1968. Marine magnetic anomalies, geomagnetic field reversals and motions of the ocean floor and continents. *J. Geophys. Res.*, 73:2119–2136.

Herron, E.M. and Hayes, D.E., 1969. A geophysical study of the Chile Ridge. *Earth Planet. Sci. Lett.*, 6:77–83.

Hess, H.M., 1962. History of ocean basins. In: A.E.J. Engel, H.L. James and B.F. Leonard (Editors), *Petrologic Studies: A Volume in Honor of A.F. Buddington.* Geol. Soc. Am., New York, N.Y., pp.599–620.

Hurley, P.M. and Rand, J.R., 1969. Pre-drift continental nuclei. *Science*, 164:1229–1242.

I.A.G.A., Commission 2, Working Group 4. International Geomagnetic Reference Field 1965.0. *J. Geophys. Res.*, 74:4407–4408.

Irving, E., 1964. *Paleomagnetism.* Wiley, New York, N.Y., 399 pp.

Irving, E., 1970. The Mid-Atlantic Ridge at 45°N. XIV. Oxidation and magnetic properties of basalt; review and discussion. *Can. J. Earth Sci.*, 7:1528–1538.

Irving, E., Park, J.K., Haggerty, S.E., Aumento, F. and Loncarevic, B., 1970. Magnetism and opaque mineralogy of basalts from the Mid-Atlantic Ridge at 45°N. *Nature*, 228:974–976.

Isacks, B. and Molnar, P., 1971. Distribution of stresses in the descending lithosphere from a global survey of focal mechanism solutions of mantle earthquakes. *Rev. Geophys.*, 9:103–174.

Isacks, B., Oliver, J. and Sykes, L.R., 1968. Seismology and the new global tectonics. *J. Geophys. Res.*, 73:5855–5899.

Karig, D.E., 1970. Ridges and basins of the Tonga-Kermadec island arc system. *J. Geophys. Res.*, 75:239–254.

Karig, D.E., 1971. Structural history of the Mariana Island arc system. *Geol. Soc. Am. Bull.*, 82:323–344.

Katsumata, M. and Sykes, L.R., 1969. Seismicity and tectonics of the Western Pacific: Izu-Mariana–Caroline and Ryukyu-Taiwan regions. *J. Geophys. Res.*, 74:5923–5948.

Kontis, A.L. and Young, G.A., 1964. Approximation of residual total magnetic intensity anomalies. *Geophysics*, 29:623–627.

Langseth, M., 1965. Techniques of measuring heat flow through the ocean floor. In: W.H.K. Lee (Editor), *Terrestrial Heat Flow. Geophys. Monogr.*, 8. Am. Geophys. Union, Washington, D. C., pp.58–77.

Larson, R.L., 1970. *Near-Bottom Studies of the East Pacific Rise Crest and Tectonics of the Mouth of the Gulf of California*. Ph.D. Thesis, Univ. California, San Diego, 164 pp.

Larson, R.L. and Spiess, F.N., 1969. East Pacific Rise crest: a near-bottom geophysical profile. *Science*, 163:69–71.

Larson, R.L., Menard, H.W. and Smith, S.M., 1968. Gulf of California: A result of ocean-floor spreading and transform faulting. *Science*, 161:781–784.

Le Pichon, X., 1968. Sea-floor spreading and continental drift. *J. Geophys. Res.*, 73:3661–3697.

Lowrie, W. and Fuller, M.D., 1971. On the alternating field demagnetization characteristics of multidomain thermoremanent magnetization in magnetite. *J. Geophys. Res.*, 76:6339–6349.

MacDonald, G.J.F., 1965. Geophysical deductions from observations of heat flow. In: W.H.K. Lee (Editor), *Terrestrial Heat Flow. Geophys. Monogr.*, 8. Am. Geophys. Union, Washington, D.C., pp.191–210.

Marshall, M. and Cox, A., 1971. Magnetism of pillow basalts and their petrology. *Geol. Soc. Am. Bull.*, 82:537–552.

Mason, R.G. and Raff, A.D., 1961. Magnetic survey off the west coast of North America, $32°N$ latitude to $42°N$ latitude. *Bull. Geol. Soc. Am.*, 72:1259–1266.

Matsushita, S., 1967. Solar quiet and lunar daily variation fields. In: S. Matsushita and W.H. Campbell (Editors), *Physics of Geomagnetic Phenomena*. Academic Press, New York, N.Y., p.321.

Matthews, D.H. and Williams, C.A., 1968. Linear magnetic anomalies in the Bay of Biscay: a qualitative interpretation. *Earth Planet. Sci. Lett.*, 4:315–320.

McKenzie, D.P. and Morgan, W.J., 1969. Evolution of triple junctions. *Nature*, 224:125–133.

McKenzie, D.P. and Parker, R.L., 1967. The North Pacific: an example of tectonics on a sphere. *Nature*, 216:1276–1280.

McKenzie, D.P. and Sclater, J.G., 1971. The evolution of the Indian Ocean since the Late Cretaceous. *Geophys. J.*, 25:437–528.

McKenzie, D.P., Davies, D. and Molnar, P., 1970. Plate tectonics of the Red Sea and East Africa. *Nature*, 226:243–248.

McManus, D.A. and Burns, R.E., 1969. Scientific report on deep sea drilling project, leg V. *Ocean Ind.*, 4(8):40–42.

Morgan, W.J., 1968. Rises, trenches, great faults and crustal blocks. *J. Geophys. Res.*, 73:1959–1982.

Morgan, W.J., Vogt, P.R. and Falls, D.F., 1969. Magnetic anomalies and sea floor spreading on the Chile Rise. *Nature*, 222:137–142.

Morley, L.W. and Larochelle, A., 1964. Paleomagnetism as a means of dating geological events. *R. Soc. Can., Spec. Publ.*, 8:512–521.

Nagata, T., 1961. *Rock Magnetism*. Maruzen, Tokyo, revised ed., 350 pp.

Nagata, T. and Kobayashi, K., 1963. Thermo-chemical remanent magnetization of rocks. *Nature*, 197:476–477.

N.B.S., 1971. *Handbook of Mathematical Functions. Natl. Bur. Stand., Appl. Math. Ser.*, 55, 9th Printing.

Néel, L., 1955. Some theoretical aspects of rock magnetism. *Adv. Phys.*, 4:191–243.

Oliver, J. and Isacks, B., 1967. Deep earthquake zones, anomalous structures in the upper mantle, and the lithosphere. *J. Geophys. Res.*, 72:4259–4275.

Opdyke, N.D., 1968. In: R.A. Phinney (Editor), *The History of the Earth's Crust.* Princeton Univ. Press, Princeton, N.J., pp.61–72.

Opdyke, N.D. and Henry, K.W., 1969. A test of the dipole hypothesis. *Earth Planet. Sci. Lett.*, 6:139–151.

Opdyke, N.D., Glass, B., Hays, J.D. and Foster, J., 1966. Paleomagnetic study of Antarctic deep-sea cores. *Science*, 154:349–357.

Ostenso, N.A., 1962. *Geophysical Investigations of the Arctic Ocean Basin.* Geophys. Polar Res. Cent., Univ. Wisconsin, Madison, Wisc., Res. Rep., 62-4, 124 pp.

Ozima, M. and Larson, E.E., 1970. Low- and high-temperature oxidation of titanomagnetite in relation to irreversible changes in the magnetic properties of submarine basalts. *J. Geophys. Res.* 75:1003–1017.

Ozima, M. and Ozima, M. 1965. Origin of thermoremanent magnetization. *J. Geophys. Res.*, 70:1363–1369.

Ozima, M. and Ozima, M., 1971. Characteristic thermomagnetic curve in submarine basalts. *J. Geophys. Res.*, 76:2051–2056.

Ozima, M., Ozima, M. and Kaneoka, I., 1968. Potassium-argon ages and magnetic properties of some dredged submarine basalts and their geophysical implications. *J. Geophys. Res.*, 73:711–723.

Page, L., 1935. *Introduction to Theoretical Physics.* Van Nostrand, New York, N.Y., 2nd ed., 661 pp.

Parkinson, W.D., 1962. The influence of continents and oceans on geomagnetic variations. *Geophys. J.*, 6:441–449.

Peterson, M.N.A., 1969. The Glomar Challenger completes Atlantic Track; scientific goals and achievements. *Ocean Ind.*, 4(5):62–67.

Peterson, M.N.A., Edgar, N.T., Cita, M., Gardner Jr., S., Gold, R., Nigrini, C. and Von der Borch, C., 1970. *Scripps Institute of Oceanology, Initial Reports of the Deep Sea Drilling Project*, 2:413–427. Govt. Print. Off., Washington, D.C.

Pitman III, W.C. and Heirtzler, J.R., 1966. Magnetic anomalies over the Pacific-Antarctic Ridge. *Science*, 154:1164–1171.

Pitman III, W.C., Herron, E.M. and Heirtzler, J.R., 1968. Magnetic anomalies in the Pacific and sea-floor spreading. *J. Geophys. Res.*, 73:2069–2085.

Pitman III, W.C. and Talwani, M., 1971. Sea floor spreading in the North Atlantic. *Geol. Soc. Am. Bull.*, 83:619–646.

Polyak, B.G. and Smirnov, Yu.B., 1968. Relationship between terrestrial heat flow and the tectonics of continents. *Geotectonics*, 205–213.

Raff, A.D., 1968. Sea-floor spreading – another rift. *J. Geophys. Res.*, 73:3699–3705.

Raff, A.D. and Mason, R.G., 1961. Magnetic survey off the west coast of North America 40°N Latitude to 52°N Latitude. *Bull. Geol. Soc. Am.*, 72:1267–1270.

Rassokho, A.I., Senchura, L.I., Demenitskaya, R.M., Karasik, A.M., Kiselev, Yu.G. and Timoshenko, N.K., 1967. Podvodnyi sredinnyi arkticheskii khrebet, i ego mesto v sisteme khrebtov severnogo ledovitogo okeana. *Dokl. Akad. Nauk, S.S.S.R.*, 172:659–662.

Richards, M.L., 1970. *A Study of Electrical Conductivity in the Earth near Peru.* Ph.D. Thesis, Univ. California, San Diego, 126 pp.

Richards, M.L., Vacquier, V. and Van Voorhis, G.D., 1967. Calculation of magnetization of uplifts from combining topographic and magnetic surveys. *Geophysics*, 32:678–701.

Rikitake, T., 1968. Earthquake prediction. *Earth-Sci. Rev.*, 4:245–282.

Rikitake, T., 1971. Electric conductivity anomaly in the Earth's crust and mantle. *Earth-Sci. Rev.*, 7:35–65.

Ringwood, A.E., 1969. Composition and evolution of the upper mantle. In: P.J. Hart (Editor), *The Earth's Crust and Upper Mantle. Geophys. Monogr.*, 13. Am. Geophys. Union, Washington, D.C., pp.1–17.

Rona, P.A., Brakl, J. and Heirtzler, J.R., 1970. Magnetic anomalies in the northeast Atlantic between the Canary and Cape Verde Islands. *J. Geophys. Res.*, 75:7412–7420.

Schmucker, U., 1970. Anomalies of geomagnetic variations in the southwestern United States. *Scripps Inst. Oceanogr. Bull.*, 13:165 pp.

Sclater, J.G. and Cox, A., 1970. Paleolatitudes from JOIDES, deep-sea sediment cores. *Nature*, 5242:934–935.

Sclater, J.G. and Francheteau, J., 1970. The implications of terrestrial heat flow observations on current tectonic and geochemical models of the crust and upper mantle of the earth. *Geophys. J.*, 20:509–542.

Sclater, J.G. and Harrison, C.G.A., 1971. The elevation of mid-ocean ridges and the evolution of the southwest Indian Ridge. *Nature*, 230:175–177.

Sclater, J.G., Anderson, R.N. and Bell, M.L., 1971. The elevation of ridges and the evolution of the central eastern Pacific. *J. Geophys. Res.*, 76:7888–7915.

Sclater, J.G., Hawkins, J.W., Mammerickx, J. and Chase, C.G., 1972. Crustal extension between the Tonga and Lau ridges: petrologic and geophysical evidence. *Geol. Soc. Am. Bull.*, 83, in press.

Sholpo, L.E., 1967. Regularities and methods of study of the magnetic viscosity of rocks. *Izv. Akad. Nauk S.S.S.R.*, 1967 (6):99–116.

Smirnov, Ya.B., 1968. The relationship between the thermal field and the structure and development of the earth's crust and upper mantle. *Geotectonics*, 5:343–352.

Spiess, F.N. and Mudie, J.D. 1970. Small scale topographic and magnetic features. In: A.E. Maxwell, (Editor), *The Sea*. Wiley, New York, N.Y., 4: 205–250.

Stacey, F.D., 1969. *Physics of the Earth*. Wiley, New York, N.Y., 324 pp.

Strangway, D.W., Larson, E.E. and Goldstein, M., 1968. A possible cause of high magnetic stability in volcanic rocks. *J. Geophys. Res.*, 73:3787–3795.

Sykes, L.R., 1968. Seismological evidence for transform faults, sea-floor spreading and continental drift. In: R.A. Phinney (Editor), *History of the Earth's Crust*. Princeton University Press, Princeton, N.J., pp.120–150.

Sykes, L.R., 1970. Seismicity of the Indian Ocean and a possible nascent island arc between Ceylon and Australia. *J. Geophys. Res.*, 75:5041–5055.

Talwani, M., 1965. Computation with the help of a digital computer of magnetic anomalies caused by bodies of arbitrary shape. *Geophysics*, 30:797–817.

Talwani, M. and Heirtzler, J.R., 1964. Computation of magnetic anomalies caused by two-dimensional structures of arbitrary shape. In: G.A. Parks (Editor), *Computers in the Mineral Industries*. *Stanford Univ. Publ. Geol. Sci.*, 9:464–480.

Talwani, M., Le Pichon, X. and Ewing, M., 1965. Crustal structure of mid-ocean ridges. 2. Computed model from gromity and seismic refraction data. *J. Geophys. Res.*, 70:341–352.

Talwani, M., Windisch, C.C. and Langseth Jr., M.G., 1971. Reykjanes Ridge Crest: a detailed geophysical study. *J. Geophys. Res.*, 76:473–517.

Théllier, E. and Théllier, O., 1959. Sur l'intensité du champ magnétique terrestre dans le passé historique et géologique. *Ann. Géophys.* 15:285.

Tracey Jr., J.I., Sutton, G.H., 1971. *Initial Reports of the Deep Sea Drilling Project*, Vol. VIII. U.S. Govt. Printing Office, Washington, D.C., 1037 pp.

Trubiatchinskii, Demenitzkaia, R.M., Karasik, A.M. and Stchelovamov, V.G., 1970. Linear magnetic anomalies in Dreik Strait. In: G.N. Petrova (Editor), *Abstracts of Papers of the VIIIth Conference on Problems of the Permanent Geomagnetic Field*. Institute of Physics of the Earth, Moscow, 241 pp.

Uyeda, S., 1962. Thermoremanent magnetism and reverse thermoremanent magnetism. In: T. Nagata (Editor), *Proceedings of the Benedum Earth Magnetism Symposium*. Univ. Pittsburgh Press, Pittsburgh, Pa., pp.87–106.

Vacquier, V. and Uyeda, S., 1967. Palaeomagnetism of nine seamounts in the western Pacific and of three volcanoes in Japan. *Bull. Earthquake Res. Inst.*, Tokyo Univ., 45:815–848.

Vacquier, V., Raff, A.D. and Warren, R.E., 1961. Horizontal displacements in the floor of the Northeastern Pacific Ocean. *Bull. Geol. Soc. Am.*, 72:1251–1258.

Van der Voo, R., 1967. The rotation of Spain: paleomagnetic evidence from the Spanish meseta. *Palaeogeogr., Palaeoclimatol., Palaeoecol.*, 3:393–416.

Van der Voo, R., 1969. Paleomagnetic evidence for the rotation of the Iberian Peninsula. *Tectonophysics*, 7:5–56.

Van Dongen, P.G., 1967. The rotation of Spain: palaeomagnetic evidence from the eastern Pyrenees. *Palaeogeogr., Palaeoclimatol., Palaeoecol.*, 3:417–432.

Vening Meinesz, F.A., 1962. Pattern of convection currents in the Earth's mantle. *K. Ned. ,Akad. Wet., Proc., Ser B*, 65:131–143.

Vestine, E.H., Lange, I., LaPorte, L. and Scott, W.E., 1947. The geomagnetic field, its description and analysis. *Carnegie Inst. Wash., Publ.*, 580:390 pp.

Vine, F.J., 1968. Magnetic anomalies associated with mid-ocean ridges. *Nature* ,199:947–949.

Vine, F.J. and Hess, H.H., 1970. Sea-floor spreading. In: A.E. Maxwell (Editor), *The Sea*. Wiley, New York, N.Y., 4:587–622.

Vine, F.J. and Matthews, D.H., 1963. Magnetic anomalies over ocean ridges. *Nature*, 199:947–949.

Vogt, P.R. and Ostenso, N.A., 1970. Magnetic and gravity profiles across the Alpha Cordillera and their relation to Arctic sea-floor spreading. *J. Geophys. Res.*, 75:4925–4937.

Vogt, P.R., Anderson, C.N., Bracey, D.R. and Schneider, E.D., 1970a. North Atlantic magnetic smooth zones. *J. Geophys. Res.*, 75:3955–3968.

Vogt, P.R., Ostenso, N.A. and Johnson, G.L., 1970b. Magnetic and bathymetric data bearing on sea-floor spreading north of Iceland. *J. Geophys. Res.*, 75:903–920.

Vogt, P.R., Anderson, C.N. and Bracey, D.R., 1971. Mesozoic magnetic anomalies, sea-floor spreading and geomagnetic reversals in the Southwestern N. Atlantic. *J. Geophys Res.*, 75:4796–4823.

Von Herzen, R.P. and Maxwell, A.E., 1959. The measurement of thermal conductivity of deep-sea sediments by the needle-probe method. *J. Geophys. Res.*, 64:1557–1563.

Wasilewski, P.J., 1968. Magnetization of ocean basalts. *J. Geomagn. Geoelectr.*, 20:29–154.

Watkins, N.D., 1968. Short period geomagnetic polarity events in deep-sea sedimentary cores. *Earth Planet. Sci. Lett.*, 4:341–349.

Watkins, N.D., 1972. A review of the development of the geomagnetic polarity time scale and a discussion of prospects for its finer definition. *Geol. Soc. Am. Bull.*, 83:551–574.

Watkins, N.D. and Richardson, A., 1968. Paleomagnetism of the Lisbon volcanics. *Geophys. J.*, 15:287–304.

Wegener, A., 1929. *Die Entstehung der Kontinente und Ozeane*. Vieweg, Braunschweig. (English Transl. by J. Biram, 246 pp., Dover Publ., Inc., New York, 1966.)

Weissel, J.K. and Hayes, D.E., 1971. Contrasting zones of sea-floor spreading south of Australia. *Nature*, 231:518–521.

White, A., in preparation. Magnetic variations across the northern Gulf of California.

Whitmarsh, R. and Jones, M., 1969. Daily variations and secular variations of the geomagnetic field from shipboard observations in the Gulf of Aden. *Geophys. J.*, 18:477–483.

Wilson, J.T., 1965. A new class of faults and their bearing upon continental drift. *Nature*, 207:343–347.

Yasui, M., Nagasaka, K., Kishii, T. and Halunen, A.J., 1968a. Terrestrial heat flow in the Okhotsk Sea (2). *Oceanogr. Mag.*, 20:73–86.

Yasui, M., Kishii, T., Watanabe, T. and Uyeda, S., 1968b. Heat flow in the Sea of Japan. In: L. Knopoff, C.L. Drake and P.J. Hart (Editors), *The Crust and Upper Mantle of the Pacific Area. Geophys. Monogr.*, 12. Am. Geophys. Union, Washington, D.C., pp.3–16.

INDEX